T0254876

Communications in Computer and Information Science 805

Commenced Publication in 2007
Founding and Former Series Editors:
Alfredo Cuzzocrea, Xiaoyong Du, Orhun Kara, Ting Liu, Dominik Ślęzak, and Xiaokang Yang

More information about this series at http://www.springer.com/series/7899

Rajnish Sharma · Archana Mantri
Sumeet Dua (Eds.)

Computing, Analytics and Networks

First International Conference, ICAN 2017
Chandigarh, India, October 27–28, 2017
Revised Selected Papers

Editors
Rajnish Sharma
Chitkara University
Chandigarh
India

Archana Mantri
Chitkara University
Chandigarh
India

Sumeet Dua
Department of Computer Science
and Electrical Engineering
Louisiana Tech University
Ruston, LA
USA

ISSN 1865-0929 ISSN 1865-0937 (electronic)
Communications in Computer and Information Science
ISBN 978-981-13-0754-6 ISBN 978-981-13-0755-3 (eBook)
https://doi.org/10.1007/978-981-13-0755-3

Library of Congress Control Number: 2018944421

Printed on acid-free paper

This Springer imprint is published by the registered company Springer Nature Singapore Pte Ltd.
part of Springer Nature
The registered company address is: 152 Beach Road, #21-01/04 Gateway East, Singapore 189721, Singapore

Preface

Chitkara University in India organized the maiden edition of the International Conference on Computing, Analytics, and Networks (ICAN 2017) during October 27–28, 2017, to address the key current issues in computer science engineering. Today's young computer scientists and researchers are heavily exploring and exploiting the area of different forms of computing techniques such as cloud, distributive, parallel, cluster, and mobile computing etc. for their research. There are similar trends in data analytics where researchers are creating various types of statistical techniques and behavioral models to handle massive data. Besides this, security is of paramount importance in today's connected world. Thus, building secure networks and ensuring privacy in big data and security in the cloud environment have become natural priorities for today's computer science researchers.

The objective of ICAN is to provide a common platform to researchers working in these significant domains so they can present their impactful research work in the areas of computing, analytics, and networks. The first edition of the conference, i.e., ICAN 2017, was organized with technical support from Springer, New Delhi, in association with British Columbia Institute of Technology, Canada, NewGen Software, New Delhi, and Student Chapters of IEEE, IETE, and CSI. Around 100 delegates from five different countries attended the conference. Four keynote talks by renowned researchers from different parts of the world, one hands-on tutorial, and paper presentations in three parallel tracks were the main highlights of this conference. The keynote talks delivered during the conference were:

- "Model Predictive Optimization for Energy Storage-Based Smart Grids" by Dr. Pao-Ann Hsiung, National Chung Cheng University, Taiwan
- "Intelligent Modelling of Moisture Sorption Isotherms in Milk Protein-Rich Extruded Snacks Prepared from Composite Flour" by Dr. A.K. Sharma, ICAR-NDRI, India
- "Using Open Clinical Data to Create an Embeddable Prediction System for Hospital Stay" by Dr. Vijay Mago, Lakehead University, Canada
- "Quantum Communication – Analyzing Fundamental Building Blocks for Quantum-Based Communication" by Dr. Joseph Spring, University of Hertfordshire, UK

The hands-on tutorial entitled "Meta-Heuristic and Hyper-Hyper-Heuristic Applications for Search-Based Software Engineering Problems" was delivered by Dr. Kamal Z. Zamli, Universiti Malaysia Pahang, Malaysia.

The conference was announced in October 2016 as a twofold event – a National Symposium on Computing, Analytics, and Networks (NCAN 2017) conducted on April 15, 2017, and the International Conference on Computing, Analytics, and Networks (ICAN 2017) conducted during October 27–28, 2017. The aim of NCAN 2017 was to provide budding researchers from India with an opportunity to present their work and obtain feedback from experts thereby preparing themselves for their more meaningful participation in ICAN 2017 later. A team of about 50 reviewers comprising leading researchers and industry professionals from all over the world reviewed the papers received for ICAN 2017. On the basis of the reviews, only 16 papers were shortlisted for oral presentation. A double-blind review process was followed and each paper was reviewed by at least three reviewers.

We are quite confident and committed to making ICAN a regular feature of Chitkara University. We look forward to receiving your quality research work for the next edition of ICAN, which will be held in 2019. In addition, Chitkara University also conducts a prominent international conference on Wireless Network and Embedded Systems (WECON). The sixth edition of WECON is to be held in November 2018. In the same year, we are also organizing ACM Compute 2018. We invite everyone working in the related areas of computer science and electronics to submit their quality research work to these conferences.

Organizing the maiden edition of an international conference with such grace and impact was challenging. We extend our sincere thanks for the tireless support and commitment of a large number of people in the background who helped us in achieving the objectives of the conference.

First and foremost, we would like to express our heartfelt gratitude to Dr. Ashok K. Chiktara, Chancellor, Chitkara University, and Dr. Madhu Chitkara, Vice Chancellor, Chitkara University, for giving us this opportunity and always motivating us to do our best to organize a high-quality conference. A major reason for the success of ICAN 2017 was Springer's association with the conference. We must thank Ms. Suvira Srivastava, Associate Editorial Director, Springer New Delhi, and her team (especially Ms. Nidhi Chandhoke) for their endless support. We thank the Technical Program Committee as well as the Advisory Committee for all their help in building a strong technical program and also helping us in getting quality research papers for the conference. We appreciate the efforts and commitment of the Reviewers Committee that resulted in the selection of very good quality papers for ICAN 2017. Thanks are also due to the local Organizing Committee from Chitkara University for their flawless handling of the conference. Most importantly, a big thanks to all the authors who submitted their research work to ICAN 2017 and to all the delegates who travelled from distant places to attend this conference. We hope to see you at the next edition of ICAN.

December 2017

Sagar Juneja
Rajnish Sharma
Archana Mantri
Sumeet Dua

Organization

International Conference on Computing, Analytics, and Networks – ICAN 2017

October 27–28, 2017 | Chitkara University, India

Organized By

Steering Committee

General Chair

Sumeet Dua	Louisiana Tech University, USA

Associate Chairs

Archana Mantri	Chitkara University, India
Rajnish Sharma	Chitkara University, India

Convener

Sagar Juneja	Chitkara University, India

Technical Program Committee

Ana Hol	University of Western Sydney, Australia
Jin Ding	Zhejiang University of Science and Technology, China
Ojaswa Sharma	IIIT Delhi, New Delhi, India
T. Nandha Kumar	University of Nottingham Malaysia Campus, Malaysia
T. S. B. Sudarshan	Amrita University, Bangalore, India
Yanhui Guo	University of Illinois, Springfield, USA

Advisory Committee

Antti Piironen	Metropolia University of Applied Sciences, Finland
B. Neelima	NMAM Institute of Technology, Karnataka, India
Igor Popov	ITMO University, Russia

James Mc Cartney	British Columbia Institute of Technology, Canada
Manisha	Infosys, India
Manish Bali	RSC India and South East Asia, India
Prakash H. Patil	IETE, Pune Center, India
P. S. V. Nataraj	IIT Bombay, India
Sadanand Gulwadi	ARM, UK
Sonia Hamnane	Digital Strategy Consultant, France
Varun Goswami	NewGen Software Technologies Limited, India

Members

Shahu Chatrapati	JNTU Hyderabad, India
Heinz-Josef Eikerling	Hochschule Osnabrueck, Germany
Manuel Frutos-Perez	University of the West of England, UK
Manoj Gulati	IIIT Delhi, India
Lakshmanan Nataraj	Mayachitra Inc., USA
Phan-Anh-Huy Nguyen	HCMUTE, Vietnam
Manjit Sandhu	GNDU, Amritsar, India
A. K. Sharma	ICAR-NDRI, Karnal, India
Michael Uelschen	University of Applied Sciences Osnabrueck, Germany
M. Vigilson Prem	Anna University, India
Rajagopalan S.	Alagappa University, India
Sandeep Arya	GJUS&T Hisar, India
Nitin Goel	CPA GLOBAL, India
Ekta Gandotra	Chitkara University, India
Veena Goswami	KIIT University, India
Sandeep Kautish	GKU, India
Neha Kishore	Chitkara University, India
Deepika Koundal	Chitkara University, India
Sumit Kushwaha	KNIT Sultanpur, India
Sujata Pandey	Amity University Uttar Pradesh, India
Deepak Prashar	LPU, India
Saurabh Sharma	GGCT, Jabalpur, India
Rajinder Sandhu	Chitkara University, India
Tamanna Sehgal	SAITM, Gurgaon, India
Atul Sharma	VastEdge, India
Rajkumar Sharma	Vikram University, Ujjain, India
Vishal Sharma	SBSSTC, Ferozepur, India
Mohd Maroof Siddiqui	Integral University, India
Jagmohan Singh	SBBS University, India
Jaiteg Singh	Chitkara University, India
Rachna Soni	DAV College, Yamunanagar, India
Koteswara Rao K.	GMRIT, India
Ashwani Kush	Kurukshetra University, India

Patrons/Sponsors/Collaborators

BRITISH COLUMBIA
INSTITUTE OF TECHNOLOGY

Student Chapter

Student Chapter

Student Chapter

Media Partner

Contents

Computing

Performance Evaluation of De-noising Techniques Using Full-Reference Image Quality Metrics

Palwinder Singh[1(✉)] and Leena Jain[2]

[1] I.K.G Punjab Technical University, Kapurthala, India
palwinder_gndu@yahoo.com
[2] Global Institute of Management and Emerging Technologies, Amritsar 143001, India
leenajain79@gmail.com

Abstract. The de-noising of digital images is crucial preprocessing step before moving toward image segmentation, representation and object recognition. It is an important to find out efficacy of filter for different noise models because filtering operation is application oriented task and performance varies according to type of noise present in images. A comparative study is made to elucidate the behavior of different spatial filtering techniques under different noise models. In this paper different types of noises like Gaussian noise, Speckle noise, Salt & Pepper noise is applied on grayscale standard image of Lenna and using spatial filtering techniques the values of full reference based image quality metrics are found and compared in tabular and graphical form. The outcome of comparative study shows that Lee, Kuan and Anisotropic Diffusion Filter worked well for Speckle noise, the Salt and Pepper noise has significantly reduced using Median and AWMF, and the Mean filter and Wiener filter works immensely efficient for reducing Gaussian noise.

Keywords: Spatial filter · Additive noise · Multiplicative noise
Image quality metrics

1 Introduction

The use of digital images is rapidly increasing in the field of education, medical diagnosis, astronomy, and manufacturing industry. There are numerous other areas in which digital images are being used. The problems associated with digital images are emerging with the increase in its application. The degradations in digital images are mainly classified by [1] as given below:

- Degradation due to non zero dimensions of picture element used in device used for imaging, which is known as spatial degradation.
- Degradation due to non zero exposure time of photosensitive material used in device used for imaging, which is known as temporal degradation.

R. Sharma et al. (Eds.): ICAN 2017, CCIS 805, pp. 3–22, 2018.
https://doi.org/10.1007/978-981-13-0755-3_1

- Degradation due to distortion of image geometry, which is known as geometrical degradations.
- Degradation due to modification in gray level of image element, which is known as noise.

Our main focus in this paper will be on degradation due to modification of gray level i.e., noise. The Number of problems in digital images can arises due to noise because the noise damages the important features of the image. Noise may occur in digital image during acquisition, transmission and retrieval process, and degrades the quality of image. The purpose of de-noising is to eliminate noise while preserving edges and other sharp transitions [2]. The removal of noise is still a difficult task for researchers because de-noising algorithms may cause blurring of edges and it may eliminate some important features of image also. The algorithm selected for de-noising depends upon the nature of the noise present in the image and the type of image to be de-noised because the de-noising algorithm used for ultrasound images may not be suitable for satellite images. Similarly the de-noising algorithm used for Gaussian noise may not be suitable for speckle noise. The de-noising is a process of restoring an image into accurate state and for this it must be known by which noise model the image has been degraded. When a degradation model is known the inverse process can be applied to restore an image back into original form. The image degradation and restoration model [2] is given as follows (Fig. 1).

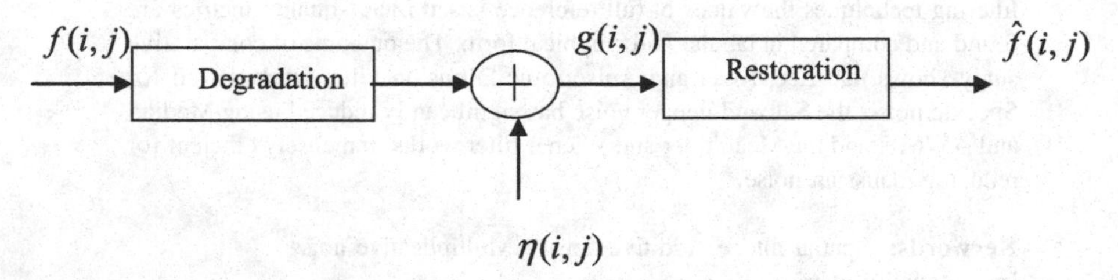

Fig. 1. Image restoration/degradation model

Where $f(i,j)$ is the original image, $g(i,j)$ is noisy image corrupted by some known degradation and additive noise $\eta(i,j)$ and $\hat{f}(i,j)$ restored image which is approximation of original image. In this paper degradation function is assumed to be identity and filtering methods will be based on noise present in the image. The spatial filtering methods Mean, Median, Adaptive Weighted Median Filter (AWMF), Wiener, Kuan, Frost and Anisotropic Diffusion filter are applied on standard image of Lenna corrupted with Gaussian noise, Speckle noise and Salt & Pepper noise respectively. The efficacy of different spatial filter has evaluated using full reference image quality metrics Root Mean Square Error (RMSE), Peak Signal to Noise Ratio (PSNR), Structural Content (SC), Average Difference (AD), Maximum Difference (MD), Normalized Correlation Coefficient (NCC), Normalized Absolute Error (NAE), Laplacian Mean Square Error (LMSE) and Structural Similarity Index (SSIM).

2 Noise Models

Image noise is unwanted signal present in captured image. Images are corrupted by noise during the process of acquisition, storing and transmission, like speckle noise is inherited in coherent imaging and occurs during image acquisition. Ultrasound, synthetic aperture radar imaging are examples of coherent imaging [4]. During transmission the noise mainly occurs in analogue channels. In this section, various noise models, their categorizations and their probability distributed function will be discussed. The noise is generally categorized as additive or multiplicative [3] which are further categorized as follow.

2.1 Gaussian Noise

The Gaussian noise is additive in nature. It is represented as in following Eq. (1).

$$g(i,j) = f(i,j) + \eta(i,j) \tag{1}$$

The probability distribution function of Gaussian noise and Gaussian noise pattern generated in matlab is given in following Fig. 2.

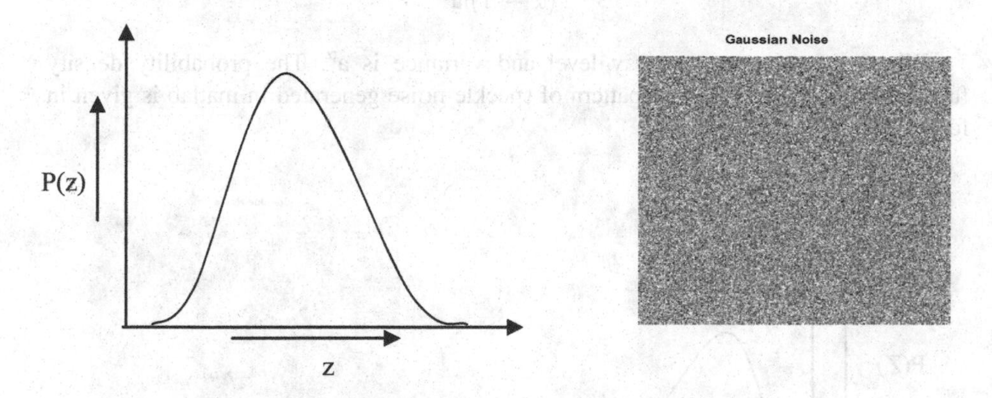

Fig. 2. Distribution function of Gaussian noise and Gaussian noise pattern.

The value of each pixel in the image having Gaussian noise will be sum of the value of pixel in reference image and the value of random Gaussian distributed noise. The Gaussian probability distribution function is given as follow in Eq. (2).

$$P(z) = \frac{1}{\sigma\sqrt{2\Pi}} e \frac{-(z - \mu)^2}{2\sigma^2} \tag{2}$$

Where μ represents mean intensity of given image, z represents the gray level and σ is the standard deviation of noise.

2.2 Speckle Noise

The Speckle noise which is multiplicative in nature emerges when coherent light illuminates the rough surface then the reflected waves from the surface contains contribution of many independent scattering areas [4]. The scattered components with relative delay of few wavelengths sums together while propagating to distant observation point and form a granular pattern and that pattern is termed as speckle. The speckle noise is intrinsic artifact of ultrasound and synthetic aperture radar imaging [5, 6]. The speckle noise is multiplicative in nature and given in following equation.

$$g(i,j) = f(i,j) * m(i,j) + n(i,j) \tag{3}$$

Equation (3) can be rewritten as

$$g(i,j) = f(i,j) * m(i,j) \tag{4}$$

The logarithmic process converts the given multiplicative equation into additive form. The probability density function for speckle noise is given below.

$$P(z) = \frac{z^{\alpha-1}}{(\alpha - 1)!a^{\alpha}} e^{-z/a} \tag{5}$$

Where z represents the gray level and variance is a^{α}. The probability density function of Speckle noise and pattern of speckle noise generated in matlab is given in following Fig. 3.

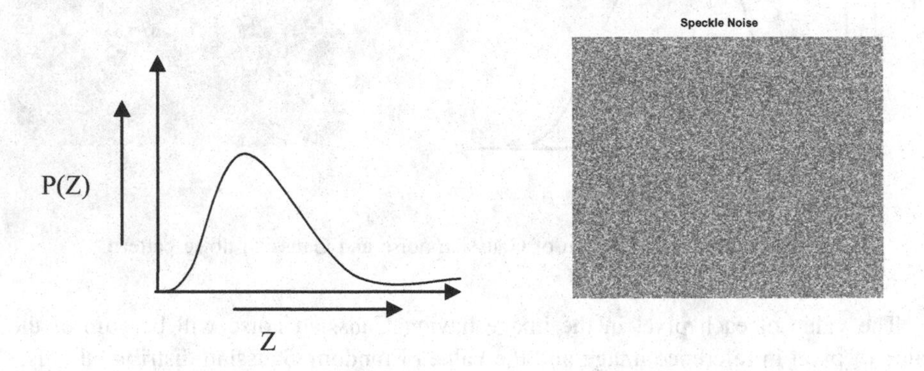

Fig. 3. Distribution function of Speckle noise and Speckle noise pattern.

2.3 Salt and Pepper Noise

The salt and pepper noise is basically an impulse noise which is caused by sudden and sharp disturbances in camera sensors or transmission in noisy channels [8]. The other type of impulse noise is random valued noise. The pixel values in image corrupted with salt and pepper noise are set to either value 'm' or 'n' and the pixels which are unaffected remain unchanged. The values 'm' and 'n' are minimum and maximum value within dynamic range. For an 8 bit image having 256 gray values, the value of salt noise is set to 255 and value of pepper noise is set to 0. An image corrupted with salt and pepper noise having dark pixels in bright region and bright pixels in dark region. The probability density function P(z) for salt & pepper noise is given as follow.

$$P(z) = \begin{cases} P_m & \text{for} \quad z = m \\ P_n & \text{for} \quad z = n \\ 0 & \text{otherwise} \end{cases} \tag{6}$$

Where 'z' represents the gray level. Once the median was the most efficient and effective filter for removing salt and pepper noise. In order to avoid these problems alternatives based on median filter like 'adaptive' or 'switching' filters [8] were proposed which we will discuss in next section. The graphical representation of probability density function of salt and pepper noise and pattern of salt and pepper noise generated in matlab is given as follow (Fig. 4).

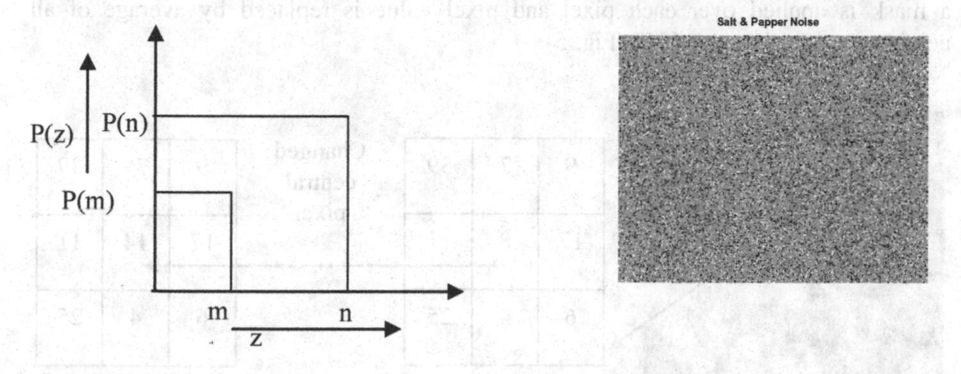

Fig. 4. Distribution function of Salt and Pepper noise and Salt and Pepper noise pattern.

3 Filtering of Digital Images

Noise may arise in digital images due to number of factors like improper sensing element, improper lightning conditions, or environment factors etc. The ultrasound images contain speckle noise during acquisition which is undesirable because as it affects the task of human interpretation and diagnosis. So before analyzing Ultrasound images, the removal of speckle noise is must.

A Significant work in spatial filtering has done by [10] using mean filter which is based on local statistical measures. The wiener filter [11] is considered as a best filter among the class of linear spatial filters. The Gaussian noise can be significantly reduced using wiener filter. In [12] Pratt also made a qualitative study of two dimensional median filters of different size and shapes. They concluded that the irrespective of size and shape of window the median filter works well for suppressing salt and pepper noise. The median filter belongs to a class of non linear spatial filters. For reducing speckle noise, L2–mean filter [13], Adaptive weighted median filter [14] and filter based on non linear diffusion [15] belongs to post formation filtering methods which works directly on original image were best. In [16] new adaptive filtering algorithm was used in which arithmetic filter was used for homogenous regions and median filter was used for edge pixels. In [23] 1990, perona and malik proposed a novel filtering using anisotropic diffusion based on the filters proposed by Lee [20], Frost [21] and Kuan [22]. These filters are based on minimizing mean square error. The use of non linear PDE methods involving anisotropic diffusion has significantly increased. In [22] noise smoothing adapts according to changes in non-stationary local mean and non-stationary local variance. In this paper efficacy of some of the existing filtering techniques like Mean, Median, AWMF, Lee, Kuan, Frost and Anisotropic diffusion filter has compared using full reference objective image quality metrics.

3.1 Mean Filter

The mean filter is used to reduce intensity variation between pixels by providing smoothness in the given digital image. It is an average filtering technique which means a mask is applied over each pixel and pixel value is replaced by average of all neighboring pixels value [11] (Fig. 5).

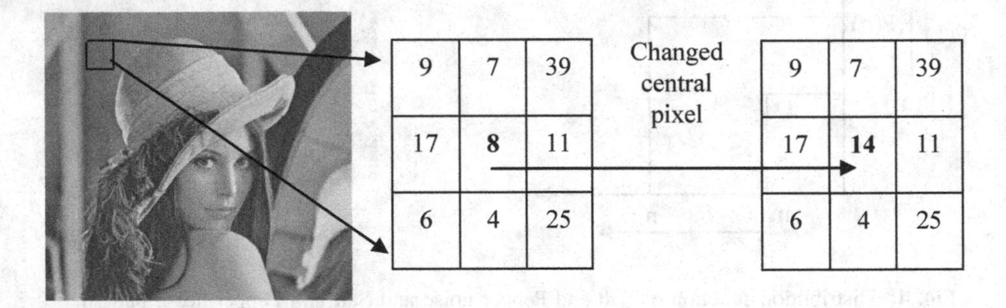

Fig. 5. Example of mean filter

Mask size begins with 3 and may increase up to finite limit but as mask size increases the blurring may also increase. The mean filter is given as follow.

$$\hat{f}(i,j) = \frac{1}{mn} \sum_{(x,y) \in S_{i,j}} g(x,y) \tag{7}$$

Where $\hat{f}(i,j)$ is de-noised image, $g(i,j)$ is degraded image and $S_{i,j}$ is the area defined by m × n size mask.

3.2 Median Filter

Median filter is the order-statistical filter which works well for salt and pepper noise but not suitable for Gaussian and speckle noise. The value of pixel (i, j) in filtered image will be median of all values given in neighborhood of (i, j) in original image. The neighborhood window may be the size of 3 × 3, 5 × 5, 7 × 7 etc. The size of window will always be in odd numbers. It also preserves the details and does not introduce artifacts in the digital image while de-noising unlike linear filters such as mean filter which may blur the edges [11, 12]. The reason is median filter is not much sensitive to extreme values unlike mean filter. The median filter is defined as follow.

$$\hat{f}(i,j) = \frac{median}{(x,y) \in w_{i,j}}(f(x,y)) \tag{8}$$

Where $f(i,j)$ is the reference image, $\hat{f}(i,j)$ is de-noised image, (i, j) is the pixel in $f(i,j)$ to be processed and $f(x,y)$ is local window. It is easy implement and easy to formulate. The algorithm for median filter is given as

Step 1: Consider each pixel in the image and select 3 × 3 size neighborhood.
Step 2: Sort all the pixels in given neighborhood into numerical order.
Step 3: Find out median value.
Step 4: Replace the given pixel with calculated median value.
Step 5: Repeat all above steps until all the pixels in the given image are replaced with median values.

3.3 Adaptive Weighted Median Filter

The adaptive median filter works on the principle that the behavior of filter applied is changes when statistical measures like mean, variance changes within image. It also changes size of neighborhood window during execution. The basic principle of algorithm remain same that we need to find median with in neighborhood window but value of given pixel will be replace with median or not, it depends upon statistical measures [17]. The algorithm for adaptive median filter is given as follow.

Step 1: Select initial window_size and maximum_window_size.
Step 2: Consider next pixel of image and repeat following steps for each and every pixel of image.
Step 3: Find out Min_value, Max_value and Median_value within given neighborhood.
Step 4: If Min_value<Median_value<Max_value
    If Min_value<Pixel_value<Max_value
  then return Pixel_value and goto step-2
    else return Median_value and goto step-2
else
ifwindow_size<maximum_window_size
  then increase size of neighborhood window and goto step-4
else return pixel_value and goto step-2

The Min_value and Max_value are minimum and maximum intensity values, generally considered as component of impulse noise. We start with finding Median_value and if it fall between Max_value and Min_value and If pixel value does not falls between Max_value and Min_value then it is considered as noise and get replaced with median value of neighborhood window, otherwise the pixel value will remain unaltered. but if median value itself is extreme then window size is increased and algorithm restart from beginning [17]. It works well for impulse noise but does not give good results for speckle and Gaussian noise.

3.4 Wiener Filter

The Wiener Filter belongs to a class of optimum stationary linear filter to filter images degraded by additive noise. There is a need to take assumption in wiener filter that signal and noise are second order stationary and performance criteria is to find filtered image such that Mean square error between degraded image and filtered image is minimum [18]. In frequency domain it is used for de-noising and de-blurring whereas in spatial domain only de-noising purpose is solved [19]. For a given degraded image g (i, j), the wiener filter is given as follow.

$$\hat{f}(i,j) = \bar{\mu} + \frac{\sigma_f^2}{\sigma_f^2 + \sigma_n^2}(g(i,j) - \bar{\mu}) \tag{9}$$

Where $\hat{f}(i,j)$ is de-noised image, $\bar{\mu}$ is local mean intensity, σ_f^2 is local variance, and σ_n^2 is variance of reference image. It is adaptively applied to noisy image which means it perform less smoothing for large variance and more smoothing for more smoothing for small variance.

3.5 Lee Filter

The Lee Filter [20] is based on local statistical measure for de-noising and for preserving edges and fine details. It is an adaptive filter which means it performs filtering if variance over an area is high otherwise filtering operation will ignored. Initially mean and variance of each pixel is derived from its local mean and variance then the estimator which minimizes the mean square error is applied to get the de-noising algorithm. The Mathematical formula for Lee filter is given as follow.

$$\hat{f}(i,j) = W(i,j)[g(i,j) - \bar{\mu}] + \bar{\mu} \tag{10}$$

Where $W(i,j) = 1 - \frac{\sigma_n^2}{\sigma_g^2}$ when $\sigma_g^2 > \sigma_n^2$ and $W(i,j) = 0$ when $\sigma_n^2 > \sigma_g^2$.

We know that σ_n^2 is variance of noisy image and σ_g^2 is variance of de-noised image. The range of W(i, j) lies between 0 and 1.

3.6 Frost Filter

In [21] a de-noising model for radar image was developed for radar images which were corrupted by multiplicative noise. Ultrasound images and SAR images are usually corrupted by multiplicative noise and standard spatial filtering techniques do not work for multiplicative speckle noise. A model for de-noising was presented to minimize mean square error for smoothing Radar images. Due to non-coherent behavior the Radar images contains multiplicative noise and standard de-noising techniques are not applicable on it. The filter is applied on spatial domain and computationally it is very economical. It is an adaptive wiener filter which convolves the pixel value within a fixed size mask. The exponential impulse response k is given as follow.

$$k = \exp\left[-MC_y(i,j)|t|\right] \tag{11}$$

Where M is filter parameter, and $|t_0|$ is distance measured from processed pixel (i, j). The filter is developed with assumption that both noise and signal are stationary. For radar images, the noise can be modeled as stationary but on a global basis the radar images are non-stationary [21]. Therefore the filter is adapted to the changes in local properties.

3.7 Kuan Filter

In [22] the de-noising of digital images from signal dependent noise is done by using filter for noise smoothing which adapts according to changes in non-stationary local mean and local variance. The filter behave likes a point processor while smoothing uncorrelated, signal dependent noise but for multiplicative noise the filter behave like lee filter with some modification which permits various estimators for local variance of image. It can be easily extend to deal with different types of other signal dependent noises. The formula for kuan filter is same as that of lee filter which is given in Eq. (11) but value of W(i, j) is changed and given as follow.

$$W(i,j) = 1 - \frac{\sigma_n^2 / \sigma_g^2}{1 + \sigma_n^2} \tag{12}$$

Where σ_n^2 is the variance of noisy image and σ_g^2 is variance of de-noised image.

3.8 Anisotropic Diffusion Filter

Anisotropic Diffusion filter uses the concept of partial differential equation to reduce noise from digital image. Idea of modeling a filter for reducing speckle noise based on diffusion filter were first proposed by perona and malik in [23], they developed model of Speckle reducing anisotropic diffusion (SRAD). It can reduce speckle noise and preserve as well as enhance edges using anisotropic diffusion which was not possible with standard spatial filtering methods [24]. It is iterative procedure in which images are considered to consist of sub regions and diffusion filter works for smoothening within the region. The work later on extended by weickert in [15] and proposed the coherence enhancing diffusion based on tensor valued diffusion filter for smoothing. The Partial differential equation proposed by perona and malik for discrete domain is given as follow.

$$I_{i,j}^{t+\Delta t} = I_{i,j}^{t+\Delta t} + \frac{\Delta t}{|S_{i,j}|} \sum_{p \in S_{i,j}} c\left(\nabla I_{(i,j),p}^{t}\right) \nabla I_{(i,j),p}^{t} \tag{13}$$

Where $I_{i,j}^{t}$ is the digital image, (i, j) denotes pixel position in 2 dimensional grid, Δt is the step size and $S_{i,j}$ is the spatial neighborhood of pixel (i, j).

4 Image Quality Metrics

The quality of image can be judged on various parameters. The applications in which the human eye is the ultimate observer the subjective quality measures are used but there are some applications in which assessment of image quality should be made automatically on the basis of automated measures without any human involvement. The subjective metrics are categorized as single stimulus in which subject evaluates the quality of test image without any reference or source image and double stimulus in which test images are evaluated in the presence of source images [26]. The results of subjective metrics can be interpreted on few scales, like average, good, best, below average, poor etc.

The objective image quality metrics can automatically predict the quality of image using some automated models without any human assistance. Moreover it can also be used for following applications [24, 25].

- It can be used to standardize an image or algorithm for enhancement, compression or restoration.
- It can be used to keep check on quality of image.
- It can be used to optimize the given algorithm.

The objective image quality metrics can be further classified as Full-reference, Reduced-reference and No-reference [28]. The study in this paper is made on Full-reference image quality metrics (Fig. 6).

4.1 Root Mean Square Error

The mean square error is the sum of the square of difference between de-noised image and reference image intensity divided by size of image. A higher value of mean square error means the difference between de-noised image and reference image is large or image has not been de-noised [27]. Mathematically mean square error is defined as follow

$$\frac{1}{MN} \sum_{i=1}^{M} \sum_{j=1}^{N} (\hat{f}(i,j) - f(i,j))^2 \tag{14}$$

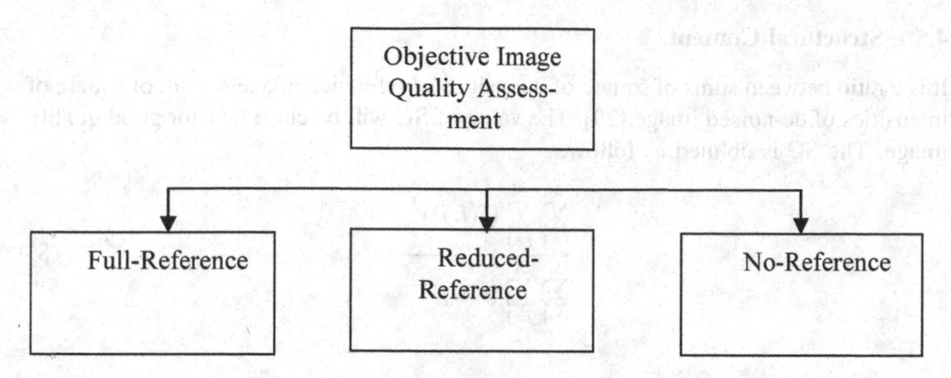

Fig. 6. The objective image quality metrics

The major drawback of MSE is its dependence on intensity scaling. The root mean square (RMSE) can be simply found by taking square root of the value of mean square error.

4.2 Peak Signal to Noise Ratio

The quality of image is measured as a ratio of maximum intensity available to the square root of mean square error [27]. For a given image f(i, j), the PSNR is mathematically defined as follow.

$$10\log\frac{S^2}{MSE} \tag{15}$$

Where S is maximum intensity and MSE is the mean square error.

4.3 Mean Absolute Error

It is the average of absolute difference between de-noised image and reference image [29]. The MAE is defined as follow.

$$\frac{1}{MN}\sum_{i=1}^{M}\sum_{j=1}^{N}\left|\hat{f}(i,j)-f(i,j)\right| \tag{16}$$

A lower value of MAE means the de-noised image is close to reference image.

4.4 Maximum Difference

It is used to measure maximum difference between de-noised image and reference image. In simple words it is used to measure the maximum error [29]. The MD is defined as follow.

$$\max\left|\hat{f}(i,j)-f(i,j)\right| \tag{17}$$

4.5 Structural Content

It is a ratio between sums of square of intensities of reference image to sum of square of intensities of de-noised image [29]. The value of SC will be close to 1 for good quality image. The SC is defined as follow.

$$\frac{\sum_{i=1}^{M}\sum_{j=1}^{N}(f(i,j))^2}{\sum_{i=1}^{M}\sum_{j=1}^{N}(\hat{f}(i,j))^2} \tag{18}$$

4.6 Normalized Absolute Error

It is used to find error prediction accuracy of the de-noised image. The SC is defined as follow [29].

$$\frac{\sum_{i=1}^{M}\sum_{j=1}^{N}\left|\hat{f}(i,j)-f(i,j)\right|}{\sum_{i=1}^{M}\sum_{j=1}^{N}\left|f(i,j)\right|} \tag{19}$$

4.7 Normalized Cross Correlation

It is a correlation function which is used to measure closeness between reference image and de-noised image [29]. It is used to measure similarity, if two images are identical images the value of NK will be 1. The NK is defined as follow.

$$\frac{\sum_{i=1}^{M}\sum_{j=1}^{N}(f(i,j))\cdot\hat{f}(i,j)}{\sum_{i=1}^{M}\sum_{j=1}^{N}(f(i,j))^2} \tag{20}$$

4.8 Laplacian Mean Square Error

Local contrast is having a crucial role in definition image quality. The LMSE is a method used for evaluating local contrast of image [29]. The LMSE is defined as follow.

$$\frac{\sum_{i=1}^{M}\sum_{j=1}^{N}(L(\hat{f}(i,j))-L(f(i,j)))^2}{\sum_{i=1}^{M}\sum_{j=1}^{N}L(f(i,j))^2} \tag{21}$$

Where

$$L(f(i,j)) = f(i+1,j) + f(i-1,j), f(i,j+1) + f(i,j-1) - 4f(i,j)$$
$$L(\hat{f}(i,j)) = \hat{f}(i+1,j) + \hat{f}(i-1,j), \hat{f}(i,j+1) + \hat{f}(i,j-1) - 4\hat{f}(i,j)$$

4.9 Structural Similarity Index

It is used to quantify structural changes which include luminance, contrast and texture of digital image. The greater value of SSIM means greater similarity between reference image and de-noised image [30]. The SSIM is defined as follow.

$$\frac{\left(2\mu_f\mu_{\hat{f}} + C_1\right) \cdot \left(2\sigma_{f\hat{f}} + C_2\right)}{\left(\mu_{\hat{f}}^2 + \mu_{\hat{f}}^2 + C_1\right) \cdot \left(\sigma_{\hat{f}}^2 + \sigma_{\hat{f}}^2 + C_2\right)} \tag{22}$$

Where μ_f is the mean intensity of reference image, $\mu_{\hat{f}}$ is mean intensity of de-noised image, σ_f^2 is variance of reference image, $\sigma_{\hat{f}}^2$ is variance of de-noised image, $\sigma_{f\hat{f}}$ is Covariance of reference and de-noised image.

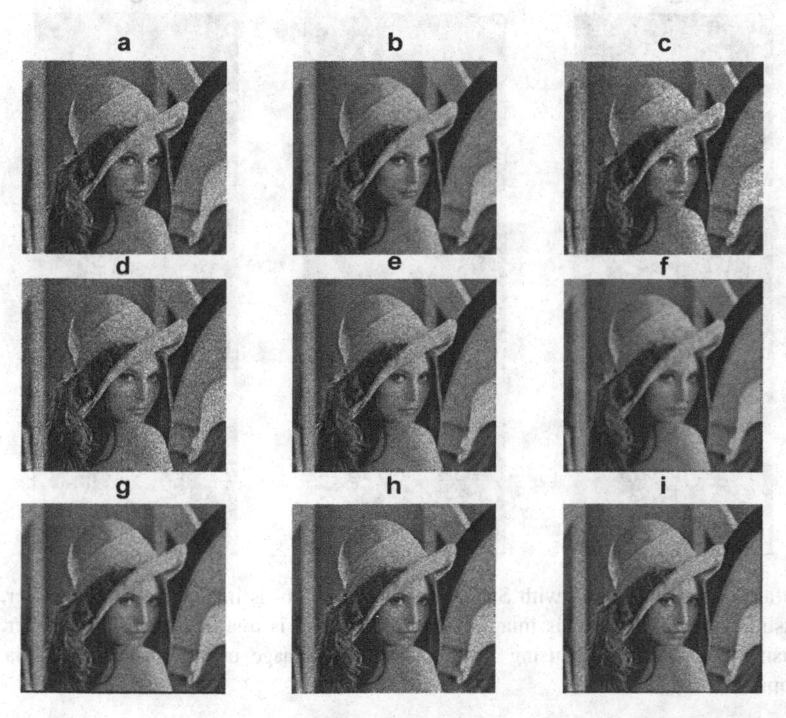

Fig. 7. 'a' is image corrupted with Speckle noise, 'b' is image filtered with Mean filter, 'c' is image filtered with Median filter, 'd' is image filtered with Adaptive_median filter, 'e' is image filtered with Wiener_filter, 'f' is image filtered with Lee filter, 'g' is image filtered with Kuan filter, 'h' is image filtered with Frost filter, 'i' is image filtered with Anisotropic_Diffussion filter.

5 Results and Discussions

The algorithms are compared using experimental evaluation, by adding speckle noise of variance 0.1, Salt & Pepper noise of variance 0.1 and Gaussian noise of mean 0 and variance 0.1 on standard grayscale test image of Lenna of size 512 × 512 and class unit 8 taken from USC-SIPI image database. The quality of filtered image is compared using full reference objective quality measures PSNR, RMSE, AD, SC, NCC, MD, LMSE, NAE and SSIM. The implementation is done on Matlab R2016B (Version 9.1). The noisy image and filtered images obtained by applying various spatial filtering techniques are given in Figs. 7, 8 and 9 where Fig. 7 shows image corrupted with speckle noise and images filtered using various spatial filtering techniques. Similarly Figs. 8 and 9 shows filtering results on image corrupted using salt and pepper noise and image corrupted using Gaussian noise. The values of objective quality metrics on noisy and filtered images are given in Tables 1, 2 and 3. Visual comparisons of objective quality metrics on noisy and filtered images are given in Figs. 10, 11, 12, 13, 14 and 15 in the form of bar charts.

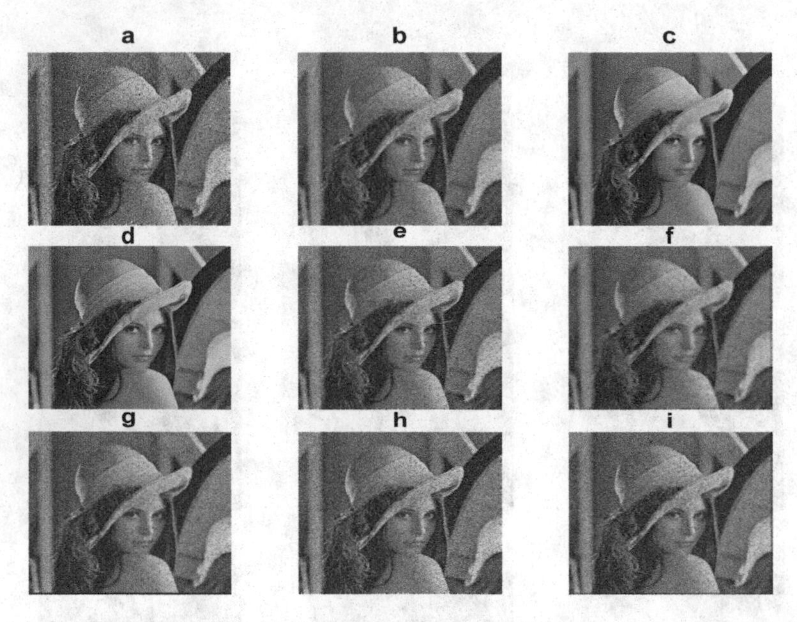

Fig. 8. 'a' is image corrupted with Salt & Pepper noise, 'b' is image using Mean filter, 'c' is image using Median filter, 'd' is image using AWMF, 'e' is image using Wiener_filter, 'f' is image using Lee filter, 'g' is using Kuan filter, 'h' is image using Frost filter, 'i' is using Anisotropic filter.

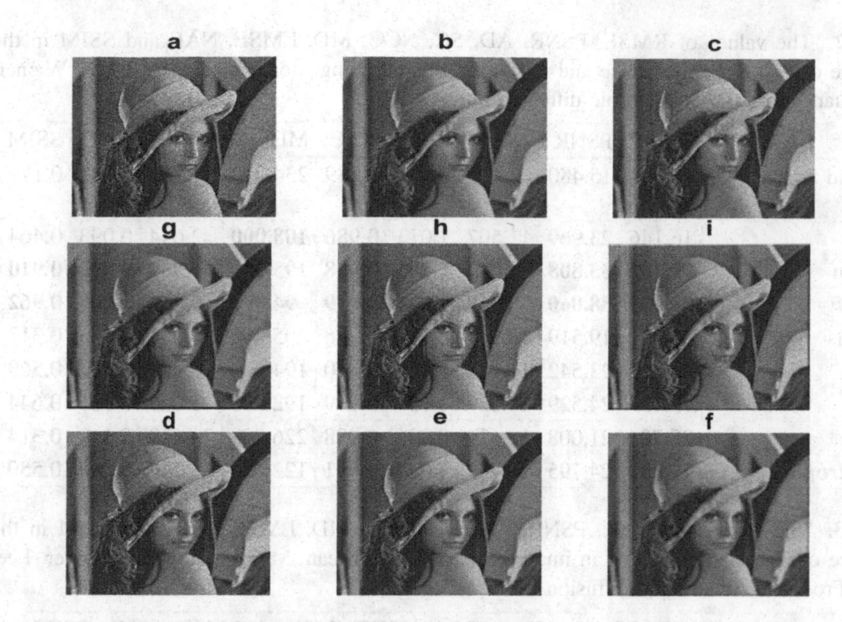

Fig. 9. 'a' is image corrupted with Gaussian noise, "b' is image using Mean filter, 'c' is image using Median filter, 'd' is image using AWMF, 'e' is image using Wiener_filter, 'f' is image using Lee filter, 'g' is using Kuan filter, 'h' is image using Frost filter, 'i' is using Anisotropic filter.

Table 1. The values of RMSE, PSNR, AD, SC, NCC, MD, LMSE, NAE and SSIM in the presence of Speckle noise and in images filtered using Mean, Median, AWMF, Wiener, Lee, Kuan, Frost and Anisotropic diffusion filters.

	RMSE	PSNR	AD	SC	NCC	MD	LMSE	NAE	SSIM
Speckle noise	40.456	15.991	33.070	0.933	0.990	124.000	82.849	0.267	0.201
Mean	14.769	24.744	11.285	1.015	0.986	**89.000**	2.433	0.091	0.500
Median	22.194	21.206	17.100	0.988	0.992	115.000	7.648	0.138	0.360
AWMF	31.728	18.102	24.973	0.980	0.982	121.000	40.225	0.201	0.237
Wiener	19.722	22.232	14.644	1.004	0.987	116.000	10.886	0.118	0.434
Lee	**11.699**	**26.768**	**8.595**	**1.003**	**0.981**	114.000	**0.952**	**0.069**	**0.670**
Kuan	14.245	25.057	**9.235**	1.032	**0.979**	194.000	**0.984**	0.074	**0.663**
Frost	21.205	21.602	16.234	**0.999**	0.988	116.000	27.008	0.131	0.350
Anisotropic	**13.521**	**25.610**	9.938	1.009	**0.981**	**114.000**	3.000	**0.071**	0.578

Table 2. The values of RMSE, PSNR, AD, SC, NCC, MD, LMSE, NAE and SSIM in the presence of Salt & Pepper noise and in images filtered using Mean, Median, AWMF, Wiener, Lee, Kuan, Frost and Anisotropic diffusion filters.

	RMSE	PSNR	AD	SC	NCC	MD	LMSE	NAE	SSIM
Salt and pepper noise	42.907	15.480	12.652	0.924	0.989	234.000	91.735	0.102	0.176
Mean	16.146	23.969	11.507	1.013	0.986	**108.000**	2.644	0.093	0.464
Median	**5.202**	**33.808**	**2.782**	**1.003**	**0.998**	195.000	**0.869**	**0.022**	**0.910**
AWMF	**3.188**	**38.060**	**0.968**	**1.001**	**0.999**	**94.000**	**0.461**	**0.008**	**0.962**
Wiener	26.951	19.519	12.985	0.987	0.986	205.000	27.041	0.105	0.312
Lee	16.960	23.542	10.140	1.024	0.980	194.000	5.950	0.082	0.569
Kuan	15.139	24.529	10.027	1.030	0.979	192.000	1.003	0.081	0.644
Frost	22.705	21.008	13.624	0.996	0.988	226.000	29.629	0.110	0.314
Anisotropic	14.682	24.795	10.402	1.027	0.981	122.000	3.226	0.084	0.550

Table 3. The values of RMSE, PSNR, AD, SC, NCC, MD, LMSE, NAE and SSIM in the presence of Gaussian noise and in images filtered using Mean, Median, AWMF, Wiener, Lee, Kuan, Frost and Anisotropic diffusion filters.

	RMSE	PSNR	AD	SC	NCC	MD	LMSE	NAE	SSIM
Gaussian noise	25.303	20.067	20.257	0.965	1.000	112.000	32.358	0.163	0.265
Mean	**9.984**	**28.145**	7.796	**1.001**	0.997	**69.000**	1.439	**0.063**	**0.642**
Median	11.710	26.760	9.222	0.998	0.997	**74.000**	2.509	0.074	0.564
AWMF	18.006	23.023	14.242	0.985	0.998	81.000	13.542	0.115	0.368
Wiener	11.257	**28.703**	8.616	**0.999**	0.997	80.000	2.899	0.069	**0.794**
Lee	**9.387**	28.680	**6.588**	1.014	**0.991**	85.000	**1.008**	**0.053**	0.759
Kuan	12.331	26.311	**7.156**	1.013	**0.989**	189.000	**0.962**	0.058	0.755
Frost	13.743	25.369	10.898	0.993	0.998	89.000	11.205	0.088	0.467
Anisotropic	10.452	27.747	7.299	1.012	0.991	105.000	1.768	0.059	0.703

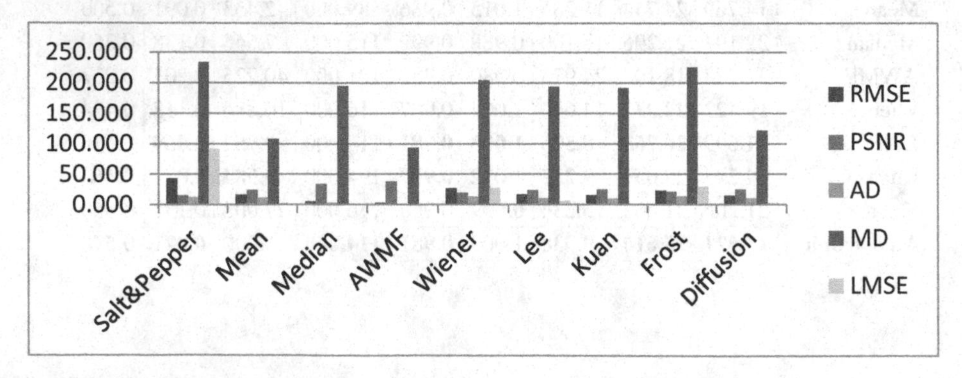

Fig. 10. Comparison of RMSE, PSNR, AD, MD and LMSE in the presence of Salt & Pepper noise and in images filtered using Mean, Median, AWMF, Wiener, Lee, Kuan, Frost and Anisotropic diffusion filters.

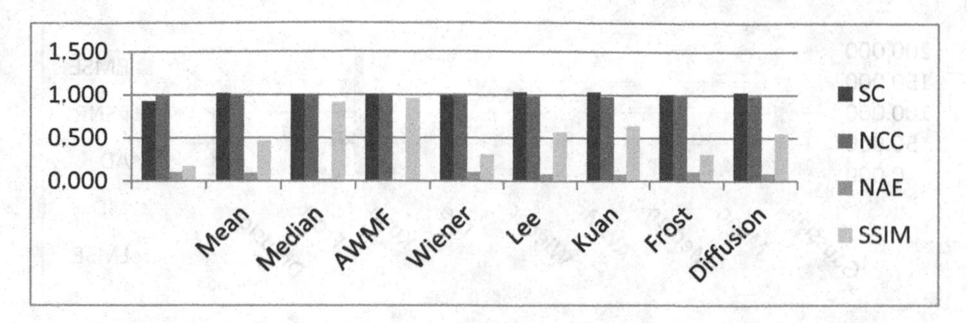

Fig. 11. Comparison of SC, NCC, NAE and SSIM in the presence of Salt & Pepper noise and in images filtered using Mean, Median, AWMF, Wiener, Lee, Kuan, Frost and Anisotropic diffusion filters.

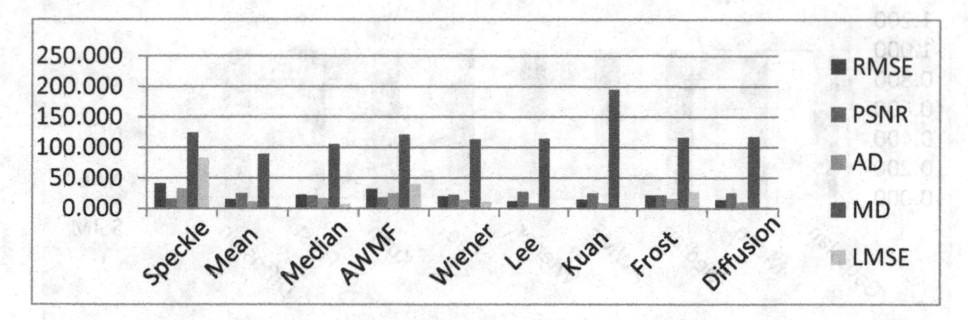

Fig. 12. Comparison of RMSE, PSNR, AD, MD and LMSE in the presence of Speckle noise and in images filtered using Mean, Median, AWMF, Wiener, Lee, Kuan, Frost and Anisotropic diffusion filters.

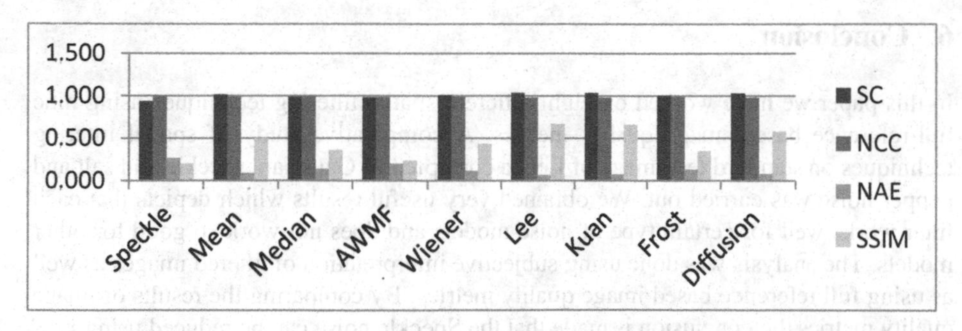

Fig. 13. Comparison of SC, NCC, NAE and SSIM in the presence of Speckle noise and in images filtered using Mean, Median, AWMF, Wiener, Lee, Kuan, Frost and Anisotropic diffusion filters.

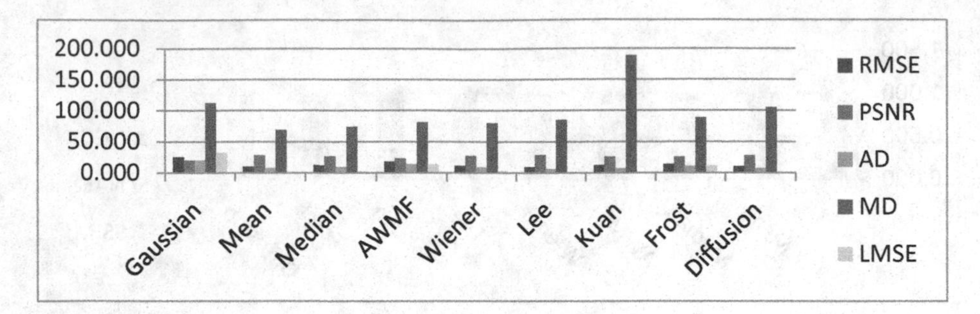

Fig. 14. Comparison of RMSE, PSNR, AD, MD and LMSE in the presence of Gaussian noise and in images filtered using Mean, Median, AWMF, Wiener, Lee, Kuan, Frost and Anisotropic diffusion filters.

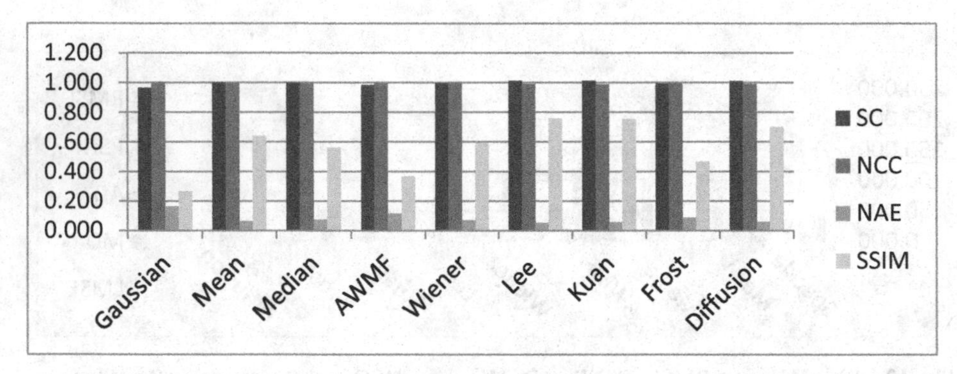

Fig. 15. Comparison of SC, NCC, NAE and SSIM in the presence of Gaussian noise and in images filtered using Mean, Median, AWMF, Wiener, Lee, Kuan, Frost and Anisotropic diffusion filters.

6 Conclusion

In this paper we have worked on eight different spatial filtering techniques using nine full-reference based image quality metrics. A comparative study of spatial filtering techniques on standard test image of Lenna corrupted of Gaussian, speckle and salt and pepper noise was carried out. We obtained very useful results which depicts that each filter works well for certain type of noise models and does not work so good for other models. The analysis was done using subjective interpretation of filtered images as well as using full reference based image quality metrics. By comparing the results of image quality metrics the conclusion is made that the Speckle noise can be reduced using Lee, Kuan and Anisotropic diffusion filter, Salt and Pepper noise can be suppressed using Median and AWMF and for Gaussian noise mean and wiener filter are immensely efficient. Although we got good results but still there is a scope of improvement in SSIM in the presence of speckle noise. One of the main future directions is to apply transform filtering in wavelet domain on images corrupted by speckle noise.

Acknowledgement. The authors express their sincere gratitude to I.K.G Punjab Technical University, kapurthala for their support and motivation.

References

1. Sonka, M., Hlavac, V., Boyle, R.: Image Processing, Analysis and Machine Vision. Thomson, Toronto (2008)
2. Gonzalez, R.C., Woods, R.E.: Digital Image Processing. Pearson Prentice Hall, Upper Saddle River (2008)
3. Pitas, I., Venetsanopoulos, A.N.: Nonlinear Digital Filters: Principles and Applications. The Springer International Series in Engineering and Computer Science. Springer, New York (1990). https://doi.org/10.1007/978-1-4757-6017-0
4. Goodman, J.W.: Some fundamental properties of speckle. J. Opt. Soc. Am. **66**(11), 1145–1150 (1976)
5. Burckhardt, C.B.: Speckle in ultrasound B-mode scans. IEEE Trans. Sonics Ultrason. **25**(1), 1–6 (1978)
6. Ma, Q., Kaplan, D.: On the statistical characteristics of log-compressed Rayleigh signals: theoretical formulation and experimental results. J. Acoust. Soc. Am. **3**, 1396–1400 (1994)
7. Motwani, M.C., Motwani, R.C., Harris, F.C., Gadiya, M.C.: Survey of image denoising techniques. In: Proceedings of Global Signal Processing, Santa Clara (2004)
8. Boncelet, C.: Image noise models. In: Bovik, A.C. (ed.) Handbook of Image and Video Processing. Academic Press, Boston (2005)
9. Tukey, J.W.: Non linear methods for smoothing data. In: Proceeding of EASCON, p. 673 (1974)
10. Jayant, N.S.: Average and median based smoothing techniques for improving digital speech quality in the presence of transmission error. IEEE Trans. Commun. **24**, 1043–1045 (1976)
11. Jain, A.K.: Fundamentals of Digital Image Processing. Prentice Hall Information and System Sciences Series. Prentice-Hall, Englewood Cliffs (1989)
12. Pratt, W.K.: Digital Image Processing, 4th edn. Wiley, New York (2007)
13. Kotropoulos, C., Pitas, I.: Optimum non linear signal detection and estimation in the presence of ultrasonic speckle. Ultrason. Imaging **14**(3), 249–275 (1992)
14. Karaman, M., Kutay, M.A., Bozdagi, G.: An adaptive speckle suppression filter for medical ultrasonic imaging. IEEE Trans. Med. Imaging **14**(2), 283–292 (1992)
15. Weickert, J.: Efficient and reliable schemes for non linear diffusion filtering. IEEE Trans. Image Process. **7**(3), 398–410 (1998)
16. Rankovic, N., Tuba, M.: Improved adaptive median filter for denoising ultrasound images. In: Proceedings of the 6th European Computing Conference, pp. 169–174 (2012)
17. Ataman, E., Wong, K.M., Aatre, B.K.: Some statistical properties of median filter. IEEE Trans. Acoust. Speech Sig. Process. **29**, 1073–1075 (1981)
18. Guan, L., Ward, R.: Restoration of randomly blurred images by the Wiener filter. IEEE Trans. Acoust. Speech Sig. Process. **37**(4), 589–592 (1989)
19. Kumar, S., Kumar, P.: Performance comparison of median and Wiener filter in image denoising. Int. J. Comput. Appl. **12**(4), 27–31 (2010)
20. Lee, J.: Digital image enhancement and noise filtering using local statistics. IEEE Trans. Pattern Anal. Mach. Intell. **2**, 165–168 (1980)
21. Frost, V.S., Stiles, J.A., Holtzman, J.C., Shanmugam, K.S.: A model for radar images and its application to adaptive digital filtering of multiplicative noise. IEEE Trans. Pattern Anal. Mach. Intell. **PAMI-4**, 157–166 (1982)

22. Kuan, D.T., Sawchuk, A.A., Chavel, P., Strand, T.C.: Adaptive noise smoothing filter for images with signal dependent noise. IEEE Trans. Pattern Anal. Mach. Intell. **PAMI-7**, 165–177 (1985)
23. Malik, J., Perona, P.: Scale space and edge detection using anisotropic diffusion. IEEE Trans. Pattern Anal. Mach. Intell. **12**, 629–639 (1990)
24. Yu, Y., Acton, S.T.: Speckle reducing anisotropic diffusion. IEEE Trans. Image Process. **11**, 1260–1270 (2002)
25. Wang, Z., Sheikh, H.R., Bovik, A.C.: Objective video quality assessment, Chap. 41. In: The Handbook of Video Databases: Design and Applications, Laboratory of Image and Video Engineering, The University of Texas, Austin, pp. 1041–1078. CRC Press (2003)
26. Pappas, T.N., Safranck, R.J.: Perceptual criteria for image quality evaluation. In: Bovik, A. C. (ed.) Handbook of Image and Video Processing. Academic Press, Boston (2000)
27. Wang, Z., Bovik, A.C.: Mean square error, love it or leave it. IEEE Sig. Process. Mag. (2009). https://doi.org/10.1109/msp2008-930648
28. Thung, K.H., Raveendran, P.: A survey of image quality measures. In: IEEE International Conference for Technical Postgraduates, pp. 1–4 (2009)
29. Eskicioglu, A.M., Fisher, P.S.: Image quality measures and their performances. IEEE Trans. Commun. **43**(12), 2959–2965 (1995)
30. Wang, Z., Bovik, A.C.: A universal image quality index. IEEE Sig. Process. Lett. **9**(3), 81–84 (2002)
31. Wang, Z., Bovik, A.C., Sheikh, H.R., Simoncelli, E.P.: Image quality assessment: from error measurement to structural similarity. IEEE Trans. Image Process. **13**(4), 600–612 (2004)

Using Open Clinical Data to Create an Embeddable Prediction System for Hospital Stay

Dillon Small[1(✉)], Fahad Wali[1(✉)], Christopher M. Gibb[1,2(✉)], and Vijay Mago[1(✉)]

[1] Lakehead University, Thunder Bay, ON P7B 5E1, Canada
{dasmall,fwali,cgibb,vmago}@lakeheadu.ca
[2] Probe Development and Biomarker Exploration, Thunder Bay Regional Health Research Institute, Thunder Bay, ON P7B 6V4, Canada

Abstract. With the ever increasing availability of open, clinical health data, there exists a deficiency of platforms to take advantage of it [1]. The global prevalence of diabetes has risen from 4.7% in 1980 to 8.5% in 2014 and continues to rise, placing an increased demand on hospital resources [2]. The management of diabetic patients within hospital can be assisted by the accurate prediction of *length of stay (LOS)* of patients. This paper introduces the use of Bayesian networks (BN) to accurately predict patient LOS in hospital. The results show the tree augmented naive BN classifier to be the most effective in predicting LOS. We believe that our model can be implemented by hospitals to more efficiently utilize hospital resources.

Keywords: Length of stay · Bayesian network · Health care system Diabetes

1 Introduction

The proportion of patients in hospital is continuing to increase, placing an added strain on medical resources. With the increasing demand for hospital services, the utilisation of resources like bed allocation and staffing have increased. The demand for hospital resources need emphasis on modelling patient flow such as patients movement throughout wards, patient length of stay (LOS) in hospital and other required resources which is considered to be vital in understanding the operation of the system and is useful in improving the functionality of the health care system.

LOS in hospital is often defined as the duration of a single patient being hospitalised. Patient LOS can be calculated by subtracting the day of admission from the day of discharge. Patient LOS in hospital is often used as an indicator of efficiency and has been an important variable for measuring the consumption of hospital resources [3,4].

R. Sharma et al. (Eds.): ICAN 2017, CCIS 805, pp. 23–33, 2018.
https://doi.org/10.1007/978-981-13-0755-3_2

A short stay unit (SSU) which provides care for patients requiring short stays have been implemented for the treatment of selected patients in certain hospitals. Their use has been found to have the possibility of reducing LOS compared to ordinary wards [5]. At the same time, patients experiencing intra-hospital transfers to a higher level of care have excess mortality and LOS [6]. By accurately predicting LOS we can correctly place certain patients in the SSU reducing their expected LOS, and reduce the number of intra-hospital transfers faced by patients mistakenly placed into an SSU, avoiding unnecessary increases in expected mortality and LOS. We perform the prediction of patient LOS using a machine learning model, *Bayesian networks (BN)*.

A BN is a probabilistic graphical diagram that represents variables and their conditional dependencies through a directed acyclic graph. The nodes represent variables and the edges represent influence or conditional dependencies between variables. BN models are well suited for enhanced decision making [7].

The objective of this paper is to predict accurate LOS of patients in hospital using a BN model. We focus our initial model on diabetic encounters. This model will help minimise intra-hospital transfers, patient risks, and LOS of patients admitted to SSUs. This minimisation in transfers and reduction of LOS can reduce the use of hospital resources and increase hospital turnover.

2 Related Works

The HbA1c measurement has an impact on patients readmission rates in the hospital [8]. The research shows that the *clinical decision support system* accurately predicts the readmission rates of patients and the cost to take care of individuals with diabetes mellitus. This information proved to be valuable in the development of strategies. The variables like LOS, readmission rates, medical speciality, age, and primary diagnosis have relationships between them. These variables have significant p values and it is observed that LOS in hospital is an important variable to predict the readmission rates of patients. BNs have been used to build numerous medical systems, but there have been a very few attempts to develop BN models in LOS in hospital:

1. Cai et al. [9] develop a predictive model for real time predictions of LOS, mortality, and readmission for hospitalised patients using electronic health records (EHRs). A BN model is developed that estimates the daily probabilities of being in the hospital, at home, or becoming deceased for a hospitalised patient. The proposed system has effective predictive power with average daily accuracy of 80% and provides a finer grained longitudinal forecast of patient status to aid in decision making at the point of care. Surprisingly, death is the most predictable outcome with a daily average accuracy of 93%.
2. Pitkaaho et al. [10] conduct a study to analyse the relationship between nurse staffing and patient LOS in an acute care unit using Bayesian dependence modelling (BDM) which belongs to the BN group. The patients acuity (measurement of the intensity of nursing care required by a patient) is the most important variable in the dependency network of nurse staffing and predicted

66% likelihood of short LOS in hospital. The study shows that flexible nurse staffing is preferable to fixed staffing to provide patients with shorter LOS in acute care units.

3. Cho et al. [11] develop a decision support intervention to predict hospital acquired pressure ulcers (HAPU) on the prevalence of ulcers and LOS in an intensive care unit (ICU), and on user adoption rates and attitudes. A BN model is used to link the EHR system to an application called Pressure Ulcers (PU) Manager. The study shows that clinical decision support intervention is an effective and sustainable strategy as the HAPU prevalence decreased and the ICU LOS shortened significantly. Results show that the use of the PU Manager has increased up to 80% in cases using the decision support system.
4. Marshall and McClean [12] develop a model for the management of the care of the elderly within hospitals, assisted by the accurate modelling of the LOS of patients in hospital. The model uses conditional phase type distributions for modelling the LOS of a group of elderly patients in hospital. The proposed model integrates the use of a Bayesian Belief Network (BBN) that describes the duration of stay in hospital. BBN permits the inclusion of additional patients information to provide better understanding of the system.
5. Marshall et al. [13] focused on modelling LOS and flow of patients especially in geriatric hospitals where elderly patients tend to have more complex needs. The proposed model predicts the LOS distribution of patients in hospital using conditional phase type distribution on a BBN. More accurate predictions for patient LOS in hospital will likely lead to improvement in management.

The main focus of these previous studies are on the results and benefits of the developed model with LOS of patients in hospital using a BN. Moreover, all the applications plan to use the hospital resources efficiently. LOS of patients in hospitals is an important variable for measuring the consumption of hospital resources. The health care system can be improved by modelling patient flow in health care systems, reducing expected LOS and use of resources. Our research accurately predicts LOS allowing health care providers to utilise resources more effectively, by placing patients into wards where they can be cared for more efficiently.

3 Methodology

3.1 Data Source

Our dataset was obtained from the UCI Machine Learning Repository and was submitted by the Centre for Clinical and Translational Research at Virginia Commonwealth University. It represents 10 years (1999–2008) of diabetic encounter data at 130 US hospitals. The dataset contains just over 100,000 data points and over 50 features denoting patient and hospital outcomes. Each datapoint is an inpatient encounter, where diabetes was entered as a diagnosis. In all cases the LOS was between 1 and 14 days, laboratory tests were performed, and medications were administered. The dataset contains attributes related to patient demographics (race, gender, age, weight, etc.), encounter information (admission type, speciality of admitting physician, LOS, etc.), test results, administered medications, and diagnoses.

3.2 Data Extraction

With the goal of developing a BN to predict the LOS of patients in hospitals, we removed a number of attributes that would not satisfy the projects criteria.

1. The attributes denoting the number of inpatient, outpatient, and emergency visits that patient had made over the past 6 months was removed due to a lack of data in the majority of cases.
2. The Discharge Disposition ID was removed as it would not be available at the time when one would want to make predictions.
3. The attributes denoting the following medications (metformin pioglitazone, acetohexamide, troglitazone, glimepiride pioglitazone, and metformin rosiglitazone) being administered were also removed as one level dominated the data with an occurrence of greater than 99.99% in the data points.
4. Weight was also removed as it was only recorded in 2 cases.
5. The samples with an admission source id corresponding to the following sources due to lack of samples. Court/Law Enforcement, Transfer from critical access hospital, Normal Delivery, Premature Delivery, Sick Baby, Extramural Birth, Transfer from hospital inpatient/same facility resulting in a separate claim, Transfer from Ambulatory Surgery Centre. For the same reasoning samples with admission type id corresponding to Newborn and Trauma Centre were also removed.
6. The 3 samples for which gender was Unknown was also removed.

In addition to removing the attributes, others were also modified. The information on diagnoses was initially encoded as ICD-9 codes in 3 different attributes. These 3 attributes were used in order to determine the rates of occurrence of all diagnoses across the data set. With this information, the 16 diagnoses with the highest rates of occurrence (5–40% of datapoints) were selected and encoded as binary variables. Finally the attributes denoting the number of lab tests performed and number of medications administered were transformed into categorical variables (split on first/second/third quartiles) as we could not use continuous variables in our network.

To learn the structure of the BN, we looked at a number of algorithms. The naive BN classifier is an effective classifier which learns the conditional probability for each attribute given the class label. It features the assumption that all attributes are conditionally independent, but despite this assumption, its predictive performance rivals state of the art classifiers. The tree-augmented naive BN classifier is an improvement over the naive BN classifier which approximates attribute interactions by imposing a Chow-Liu tree structure over the naive BN classifier structure [14]. It features increased accuracy over the naive

BN classifier. Max-Min Parents and Children is a forward selection technique for neighbourhood detection, maximising the minimum association measure with all subsets of nodes from previous iterations. It can be used to detect the undirected structure of the BN. Using the undirected skeleton network, greedy algorithms were used to search the space of directed graphs. We use Hill-Climbing, and Tabu Search, a modified hill-climbing algorithm which can escape local optima [15,16]. We also looked at Incremental Association, based on the Markov detection algorithm [17], however it was only able to return a partially directed network structure, so it was not used further. Incremental Association uses two phases, a forward phase where all variables that belong in the Markov Blanket and some false positives are added, followed by a backward phase where false positives are removed.

Using the Max-Min Parents and Children Algorithm, we were able to detect a number of isolated nodes in the network. We removed these isolated nodes from the dataset to avoid including connections with these nodes within the Naive BN Classifiers. Two of the attributes denoting diagnoses (ICD-9 codes 403 and 414) as well as attributes denoting a change in diabetic medications, and the presence of diabetic medications were removed.

After computing the tree-augmented BN, all arcs were evaluated using Pearson's chi-squared test χ^2. Two arcs, between LOS in hospital and Cardiac Dysrhythmias, as well as LOS in hospital and Chronic Airway Obstruction showed p-values of 0.242 and 0.156, respectively. As such, we could not reject the null hypothesis in these two cases and those attributes were removed from the dataset as they showed no significant influence elsewhere. All other arcs showed p-values <0.01. The attributes comprising the processed dataset can be seen below in Table 1.

Table 1. List of variables and their descriptions in the processed dataset.

Variable name	Description and values
Race	Values: Caucasian, Asian, African American, Hispanic, and other
Gender	Values: male, female, and unknown/invalid
Age	Grouped in 10-year intervals: $[0, 10), [10, 20), ..., [90, 100)$
Admission type	Integer identifier corresponding to 9 distinct values, for example, emergency, urgent, elective, newborn, and not available
Admission source	Integer identifier corresponding to 21 distinct values, for example, physician referral, emergency room, and transfer from a hospital
Length of stay in hospital	Grouped in intervals: $(0, 3], (3, 15]$; number of days between admission and discharge

(*continued*)

Table 1. (*continued*)

Variable name	Description and values
Number of lab procedures	Grouped in intervals: $(0, 31], (31, 44], (44, 57], (57, 150]$; number of lab tests performed during the encounter
Number of procedures	Number of procedures (other than lab tests) performed during the encounter
Number of medications	Grouped in intervals: $(0, 10], (10, 15], (15, 20], (20, 100]$; number of distinct generic names administered during the encounter
Number of diagnoses	Number of diagnoses entered to the system
Glucose serum test result	Indicates the range of the result or if the test was not taken. Values: >200, >300, normal, and none if not measured
A1c test result	Indicates the range of the result or if the test was not taken. Values: >8 if the result was greater than 8%, >7 if the result was greater than 7% but less than 8%, normal if the result was less than 7%, and none if not measured
14 features for diagnoses	Values: 0, 1 denoting presence of 16 different diagnoses

4 Results and Discussion

Here we consider the performance of the BNs in prediction of LOS. The BNs were trained on the processed dataset described in previous Sect. 3. The dataset was split into training and testing sets; randomly selecting 70% of cases for training and the remaining 30% for testing. Prediction accuracy was measured using the area under the curve (AUC) under receiver operating characteristic (ROC) curve. The ROC curve is generated by plotting the true positive rate against the false positive rate for various thresholds. The ROC AUC ranges between 0 and 1, with an uninformative model yielding a value of 0.5, and a value of 1 corresponding to perfect performance.

We tested three BNs for predictive accuracy. The first used the tree-augmented naive BN classifier. The second and third used max-min parents and children to detect the skeleton network, followed by searching the restricted space using hill-climbing and tabu search. Using the tree-augmented naive BN classifier, we computed an AUC of (0.782). With max-min hill-climbing and max-min tabu search, we computed AUC's of (0.657) and (0.737), respectively. The ROC plots for the three models are shown in Fig. 1.

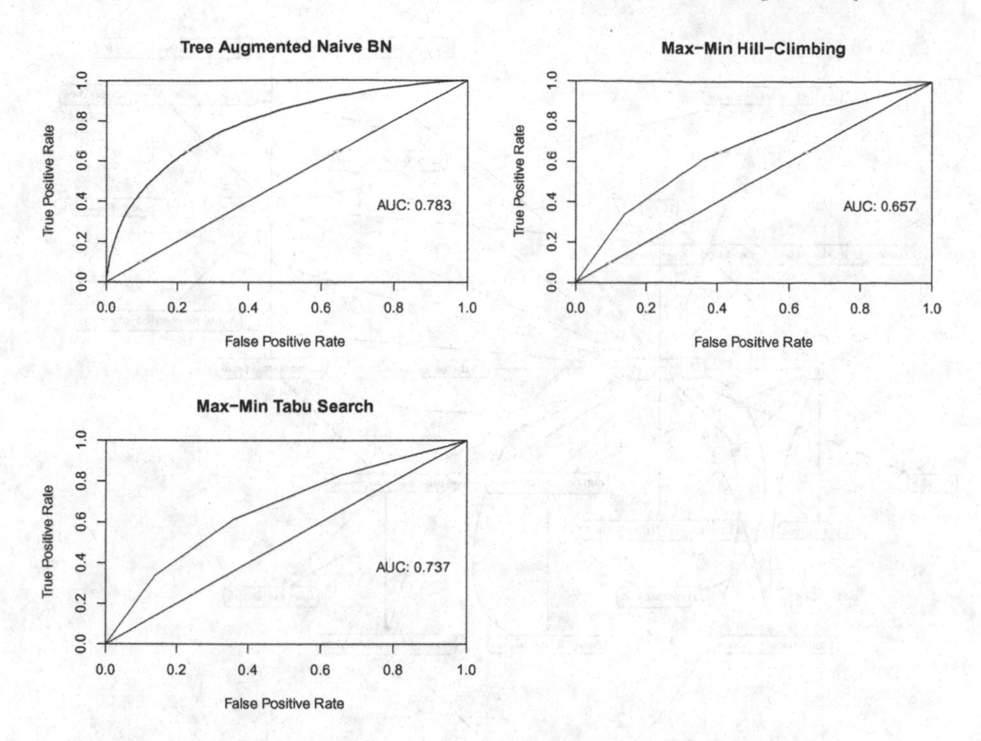

Fig. 1. ROC plots for the tree-augmented naive BN classifier, max-min hill-climbing, and max-min tabu search BNs. The area under the curve represents discrimination (accuracy of the model), the ability of the model to classify encounters between the LOS durations.

It is important to note that while the tree augmented naive BN classifier showed the highest level of predictive accuracy, it is not the true structure of the network. As can be seen below in Fig. 2, the tree augmented naive BN classifier assumes a conditional dependency between the explanatory variable and every other variable in the network. We must not draw conclusions about which attributes affect the expected LOS from the tree augmented naive BN classifier.

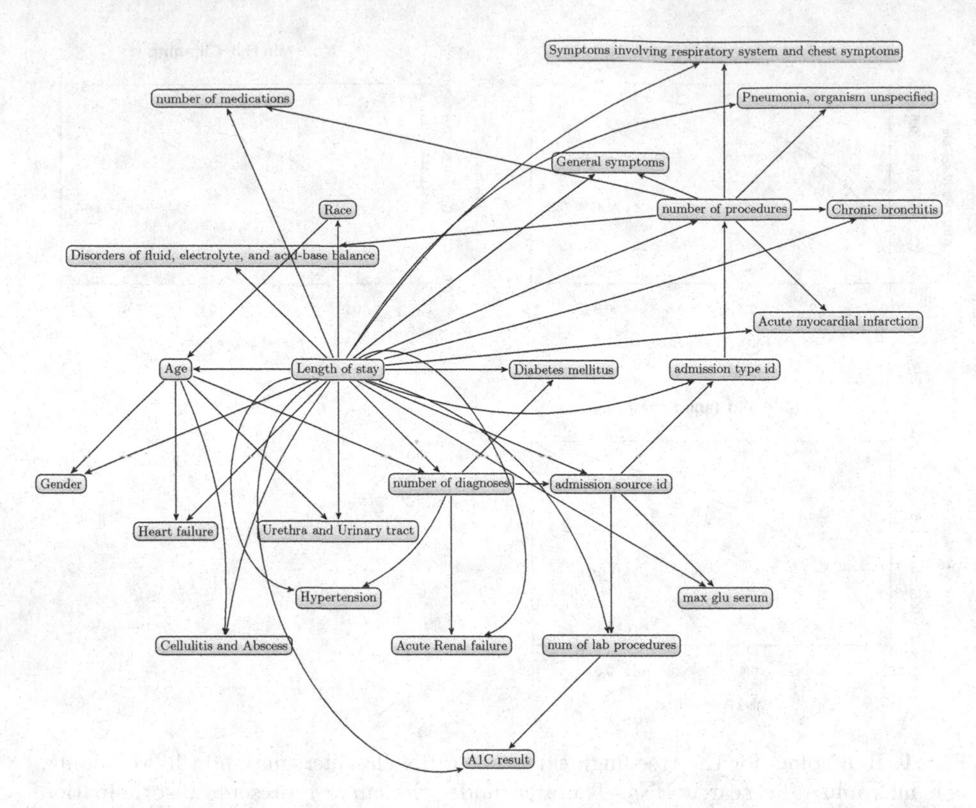

Fig. 2. Shows the directed structure of the tree augmented BN, with edges representing conditional dependencies between nodes.

Despite the lower accuracy of the BN generated using max-min tabu search, it provides a more accurate insight into the true structure of the causal relationships affecting LOS. We can draw more conclusions about how attributes contribute to a patient's LOS from the structure of the max-min tabu search BN shown below in Fig. 3. For example, diagnoses of hypertension and heart failure are the most influential diagnoses on LOS. Similar conclusions can also be drawn from the structure of the max-min hill-climbing BN shown in Fig. 4. The max-min hill-climbing BN, however, represents a less truthful causal structure, as it is unable to escape from local optima in the search space unlike max-min tabu search.

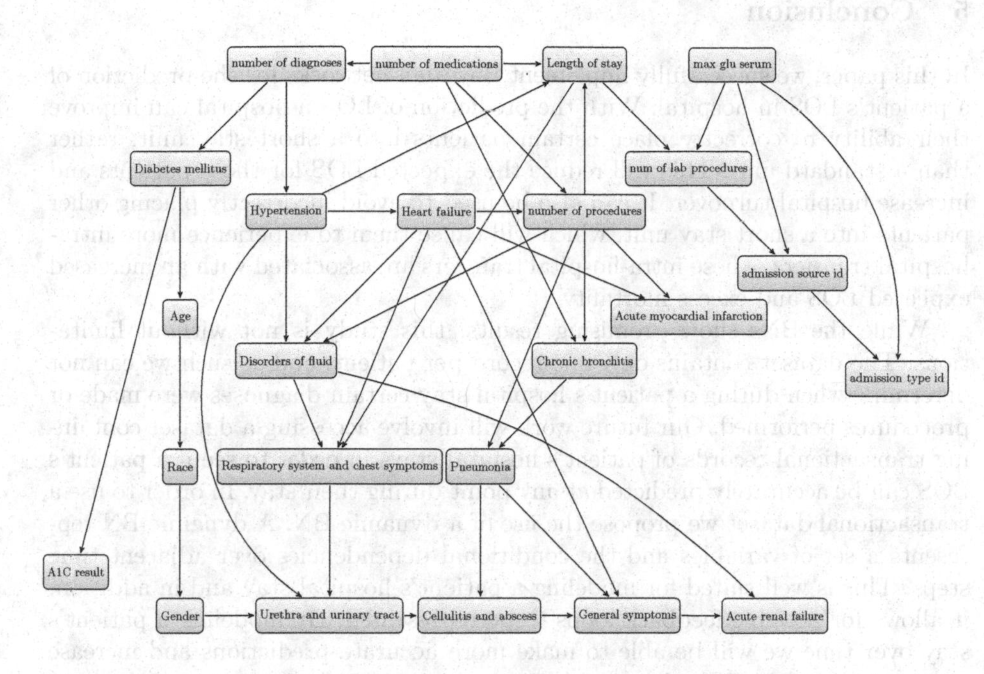

Fig. 3. Shows the directed structure of the BN found using Max-Min Tabu Search.

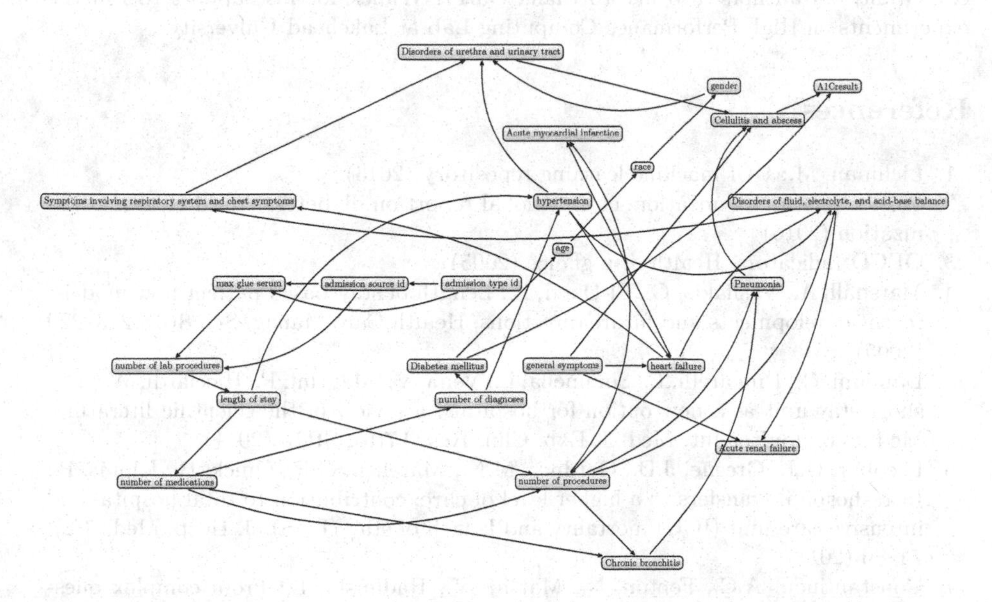

Fig. 4. Shows the directed structure of the BN found using Max-Min Hill-Climbing.

5 Conclusion

In this paper, we successfully implement Bayesian networks for the prediction of a patient's LOS in hospital. With the prediction of LOS a hospital can improve their ability to correctly place certain patients into a short stay unit, rather than a standard unit. This will reduce the expected LOS for those patients and increase hospital turnover. It can also be used to avoid incorrectly placing other patients into a short stay unit, which will cause them to experience more intra-hospital transfers. These intra-hospital transfers are associated with an increased expected LOS and excess mortality.

While the BNs show promising results, this study is not without limitations. The dataset contains only one record per patient, and as such we can not determine when during a patient's hospital stay certain diagnoses were made or procedures performed. Our future work will involve accessing a dataset containing transactional records of patient's hospital stays in order to see if a patient's LOS can be accurately predicted at any point during their stay. In order to use a transactional dataset we propose the use of a dynamic BN. A dynamic BN represents a set of variables and the conditional dependencies over adjacent time steps. This is well suited for modeling a patient's hospital stay and in addition, it allows for possible feedback loops to be represented. By modeling a patient's stay over time we will be able to make more accurate predictions and increase the models usefulness for hospitals.

Acknowledgement. The work is partially supported by Dr. Mago's NSERC Discovery Grant. All authors also like to thank Darryl Willick for his support to run the experiments on High Performance Computing Lab at Lakehead University.

References

1. Lichman, M.: UCI machine learning repository (2013)
2. World Health Organization, et al.: Global report on diabetes. World Health Organization (2016)
3. OECD Indicators. Health at a glance (2005)
4. Marshall, A., Vasilakis, C., El-Darzi, E.: Length of stay-based patient flow models: recent developments and future directions. Health Care Manag. Sci. **8**(3), 213–220 (2005)
5. Damiani, G., Pinnarelli, L., Sommella, L., Vena, V., Magrini, P., Ricciardi, W.: The short stay unit as a new option for hospitals: a review of the scientific literature. Med. Sci. Monitor Int. Med. J. Exp. Clin. Res. **17**(6), SR15 (2011)
6. Escobar, G.J., Greene, J.D., Gardner, M.N., Marelich, G.P., Quick, B., Kipnis, P.: Intra-hospital transfers to a higher level of care: contribution to total hospital and intensive care unit (ICU) mortality and length of stay (LOS). J. Hosp. Med. **6**(2), 74–80 (2011)
7. Constantinou, A.C., Fenton, N., Marsh, W., Radlinski, L.: From complex questionnaire and interviewing data to intelligent Bayesian network models for medical decision support. Artif. Intell. Med. **67**, 75–93 (2016)

8. Strack, B., DeShazo, J.P., Gennings, C., Olmo, J.L., Ventura, S., Cios, K.J., Clore, J.N.: Impact of HbA1c measurement on hospital readmission rates: analysis of 70,000 clinical database patient records. BioMed. Res. Int. **2014**, 1–11 (2014)
9. Cai, X., Perez-Concha, O., Coiera, E., Martin-Sanchez, F., Day, R., Roffe, D., Gallego, B.: Real-time prediction of mortality, readmission, and length of stay using electronic health record data. J. Am. Med. Inform. Assoc. **23**(3), 553–561 (2015)
10. Pitkäaho, T., Partanen, P., Miettinen, M., Vehviläinen-Julkunen, K.: Non-linear relationships between nurse staffing and patients? Length of stay in acute care units: Bayesian dependence modelling. J. Adv. Nurs. **71**(2), 458–473 (2015)
11. Cho, I., Park, I., Kim, E., Lee, E., Bates, D.W.: Using EHR data to predict hospital-acquired pressure ulcers: a prospective study of a Bayesian network model. Int. J. Med. Inform. **82**(11), 1059–1067 (2013)
12. Marshall, A.H., McClean, S.I.: Conditional phase-type distributions for modelling patient length of stay in hospital. Int. Trans. Oper. Res. **10**(6), 565–576 (2003)
13. Marshall, A.H., McClean, S.I., Shapcott, C.M., Millard, P.H.: Modelling patient duration of stay to facilitate resource management of geriatric hospitals. Health Care Manag. Sci. **5**(4), 313–319 (2002)
14. Friedman, N., Geiger, D., Goldszmidt, M.: Bayesian network classifiers. Mach. Learn. **29**(2–3), 131–163 (1997)
15. Tsamardinos, I., Brown, L.E., Aliferis, C.F.: The max-min hill-climbing Bayesian network structure learning algorithm. Mach. Learn. **65**(1), 31–78 (2006)
16. Tsamardinos, I., Aliferis, C.F., Statnikov, A.: Time and sample efficient discovery of Markov blankets and direct causal relations. In: Proceedings of the Ninth ACM SIGKDD International Conference on Knowledge Discovery and Data Mining, pp. 673–678. ACM (2003)
17. Tsamardinos, I., Aliferis, C.F., Statnikov, A.R., Statnikov, E.: Algorithms for large scale Markov blanket discovery. In: FLAIRS Conference, vol. 2, pp. 376–380 (2003)

Smart Toll Collection Using Automatic License Plate Recognition Techniques

S. Mahalakshmi[1(✉)] and R. Sendhil Kumar[2]

[1] BMSIT&M, Bangalore, Karnataka, India
maha.shanmugam@bmsit.in
[2] VIT University, Vellore, Tamil Nadu, India
rsenthilkumar@vit.ac.in

Abstract. The growing affluence of urban India has made the ownership of vehicles a necessity. This has resulted in an unexpected civic problem - that of traffic control and vehicle identification. Parking areas have become overstressed due to the growing numbers of vehicles on the roads today. The Automatic Number Plate Recognition System (ANPR) plays an important role in addressing these issues as its application ranges from parking admission to monitoring urban traffic and to tracking automobile thefts. There are numerous ANPR systems available today which are based on different methodologies. In this paper, we attempt to review the various techniques and their usage. The ANPR system has been implemented using OCR.

Keywords: Automatic license plate recognition · Hough line transform OCR

1 Introduction

Automatic License Plate Recognition (ALPR) is a technology which automatically recognizes number from vehicles license number plate. Vehicles license number is a unique identity as no two vehicles can have same license number. So the recognition of this unique ID is very important as it has many uses [1]. ALPR can be helpful in the case if any vehicle is involved in some crime such as accident, burglary, theft, violation of any traffic rules etc. By the recognition of license number, the criminal using that vehicle can be identified. It is also used by military for surveillance [2]. ALPR is an example of Optical character recognition (OCR). OCR is technique in which all the handwritten or printed is converted into editable form. In ALPR, the characters in the image of the number plate are written in a text file which can be edited. Thus the unique number of the vehicle is recognized by a machine which was designed by humans or computer. In this paper, identification of Indian number plate is proposed which is based on OCR [3]. Indian number plate consists of 10 characters of English language and numbers. Isolated images of all the alphabets of English language i.e. from A–Z and numerals from 0–9 are taken as templates. Then the characters on the number plate are segmented and are matched with the template created. The closest match is then recognized as the printed character and is written in a text file.

R. Sharma et al. (Eds.): ICAN 2017, CCIS 805, pp. 34–41, 2018.
https://doi.org/10.1007/978-981-13-0755-3_3

2 Existing System

Usually the car plates appear in different types of character styles, either single or double row, different sizes, spacing and character counts. Due to such kind of variations even localizing or detecting these plates a difficult problem. The problem of localization will be much harder during night time due to poor lighting conditions. Sometimes, edge geometrical feature is being used for detecting the license plates. The edge part is got using Difference of Gaussian operation followed by Sobel vertical edge mask. Before that, gamma correction is applied to the image to increase the chances of detection of edges [4]. After this we apply morphological operations to get the plate region candidates. Using these regions, along with the edge image, we calculate geometrical features of these regions and use rule-based classifier to identify the true plate region exactly.

There are a number of possible difficulties that the software must be able to cope with. These include:

- Poor image resolution, usually because the plate is too far away but sometimes resulting from the use of a low-quality camera.
- Bad images particularly blur.
- Poor lighting and low contrast due to overexposure, reflection or shadows.
- An object obscuring (part of) the plate, quite often a tow bar, or dirt on the plate.
- A different font, popular for vanity plates (some countries do not allow such plates, eliminating the problem).
- Lack of coordination between countries or states. Two cars from different countries or states can have the same number but different design of the plate.

3 Proposed System

The proposed system comprises of the following stages. The flowchart in Fig. 1 depicts the various image processing stages that finally produce image objects to the edge detection and OCR.

3.1 Implementation

Insert Image: First, the image is captured from the camera and is loaded for processing.

Processing: In processing there are 2 stages Hue, Saturation and Brightness. Hue: Degree to which a color can be described as similar or different from color that are defined Saturation: Expression for relative bandwidth of the visible output from a light source. Brightness: Relative expression intensity of energy output of a visible light source.

Grey Scale Conversion: Grey scale Conversion is a process in which the value of each pixel is a single sample that is, it carries only intensity information. Images of this sort, also known as black-and-white, are composed exclusively of shades of grey, varying from black at the weakest intensity to white at the strongest.

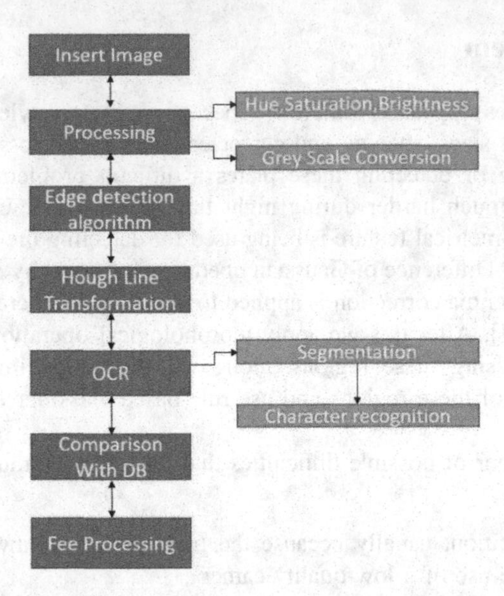

Fig. 1. Flow chart for toll collection

Edge Detection: Edge detection is an image processing technique for finding the boundaries of objects within images. It works by detecting discontinuities in brightness. Edge detection is used for image segmentation and data extraction in areas such as image processing, computer vision, and machine vision.

License plate detection module is further divided into the following subtasks.

1. Identifying edges using canny detector
2. Hough Transformation
3. Canny Edge detection

The canny detection algorithm runs in five steps as mention:

i. Smoothing: Remove the noise by blurring.
ii. Finding gradients: Where the gradients of the image have large magnitudes those edges are marked.
iii. Non-maximum suppression: Only local maxima are considered to be the edges.
iv. Double threshold: Potential edges are determined by fixing the threshold, which in our case is 0.5
v. Edge tracking by hysteresis: The end edges are determined by deleting all edges that are not connected to a very true (strong) edge as in Fig. 2(b)

All the real edges in the picture are considered even some of the edges in the background, like edges of tree or fence are also detected and we get an edge map as shown in Fig. 2(b). By considering these edges number plate of the vehicle is extracted by using Hough transformation.

Fig. 2. (a) Captured image edge detection (b) Identified edges using canny

Hough Line Transformation

Hough transform is a feature extraction technique used in image analysis, computer vision and digital image processing. It is a popular technique to detect any shape. It can detect the shape even if it is broken or distorted a little bit [4].

So using Hough transformation all the strong edges vertically and horizontally in the image is identified as show in the Fig. 3(a). When all the strong edges are identified then vertical edges are differentiated from horizontal edges as shown in the Fig. 3(b). Now using Euclidean distance all the vertical edges are grouped and checked which of the two edges have same or almost same x & y coordinates (starting and ending points), are identified.

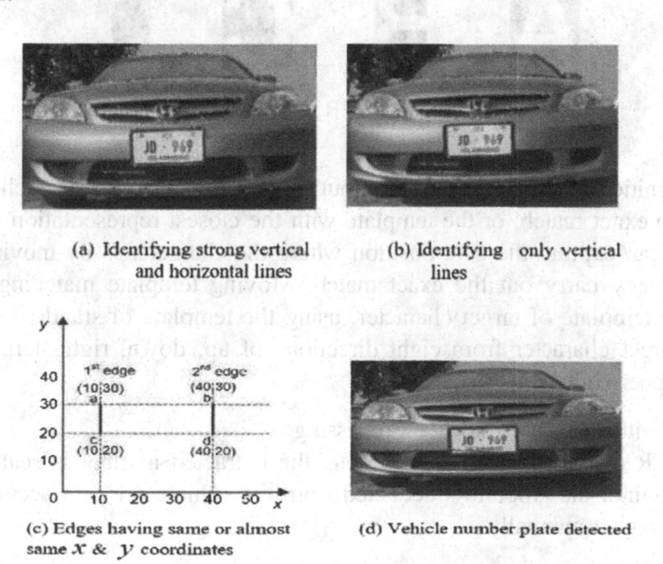

Fig. 3. Vehicle number plate detection

There are two points for each edge. For 1st edge point 'a' (10,30) and point 'c' (10,20), for 2nd edge point 'b' (40,30) and 'd' (40,20) as shown in Fig. 3(c). Horizontal dotted line shows that point a & b have same height on y-axis and in the same way horizontal dotted line shows that point c & d also lies on same height regarding y-axis. So considering these two edges and discarding all other edges, point a is joined with point b and in the same way point c is joined with point d, which yields vehicle number plate identification as shown in Fig. 3(d).

OCR (Optical Character Recognition)
It is the recognition of printed or written text characters by a computer. This involves photo scanning of text character by character analysis of the scanned-in image, and then translation of character image into character codes such ASCII.

It works by pixel-by-pixel comparison of the image and the template for each possible displacement of the template [5]. This process involves the use of a database of characters or templates. There exists a template for all possible input characters. Templates are created for each of the alphanumeric characters (from A–Z and 0–9) using 'Regular' font style. Figure 4 shows the templates for few of the alphanumeric characters.

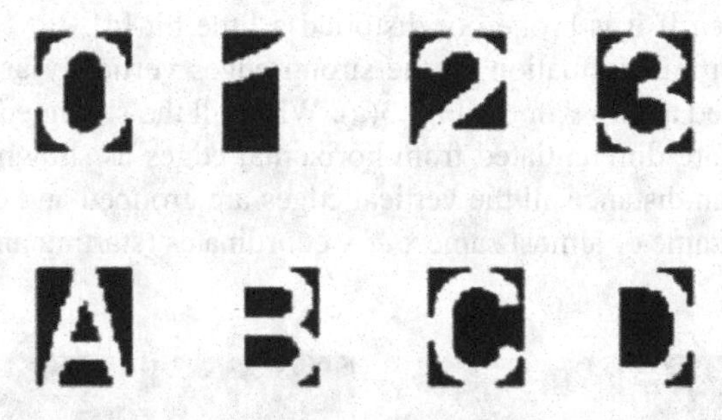

Fig. 4. OCR templates

For recognition to occur, the current input character is compared to each template to find either an exact match, or the template with the closest representation of the input character. It can capture the best position where the character is by moving standard template; thereby carry out the exact match. Moving template matching method is based on the template of target character, using the template of standard character to match the target character from eight directions of up, down, right, left, upper left, lower left, upper right, lower right [6].

Comparison with Database and Fee Processing
Once the OCR recognizes the license plate, the extracted number is searched in the database, if found, the amount is deducted from the vehicle owner's account, or else, they have to pay it manually.

4 Experimental Results

The algorithm was tested using different license plates having various background conditions, light condition and image quality. Some of the output results are shown below (Fig. 5).

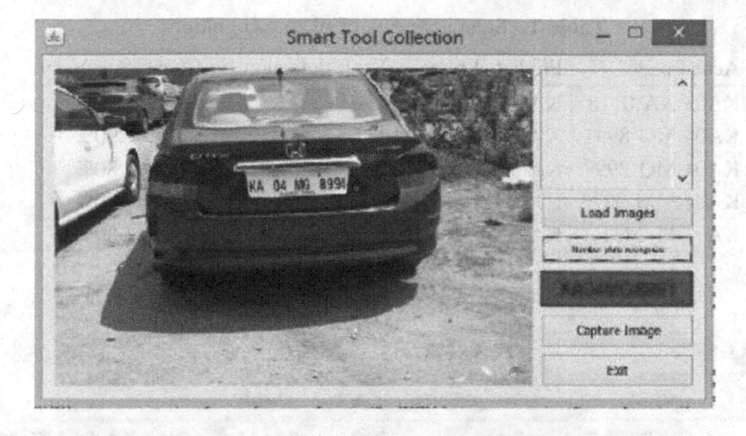

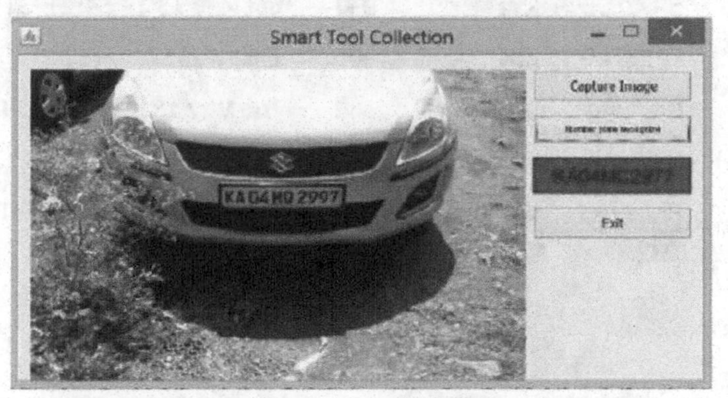

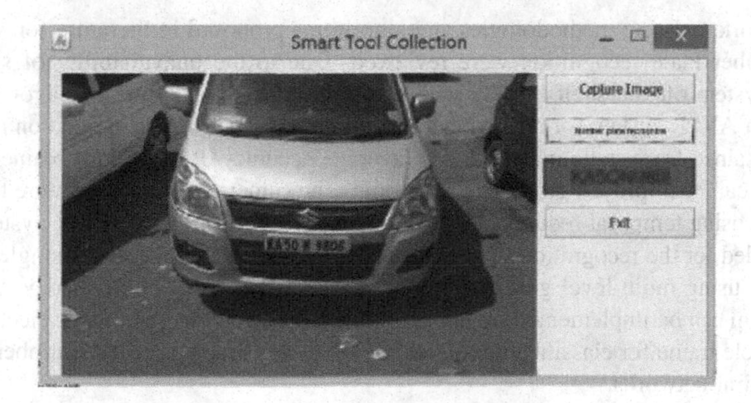

Fig. 5. Sample output after applying ALPR techniques

The results of OCR for character recognition on some of the Indian number plates taken from static images are shown in Table 1 (Fig. 6).

Table 1. Sample results of plate recognition

Actual plate	Predicted plate	Fault detection (in Char)	Accuracy
KA03 NA 0118	KA03 NA 0118	0	100%
KA04 MG 8991	XA04 MG 8991	1	90%
KA04 MQ 2997	KA04 MC 2977	2	80%
KA04 MM 3223	II04 MM 3223	2	80%
KA53 M 2431	KAI3 N 2431	2	80%

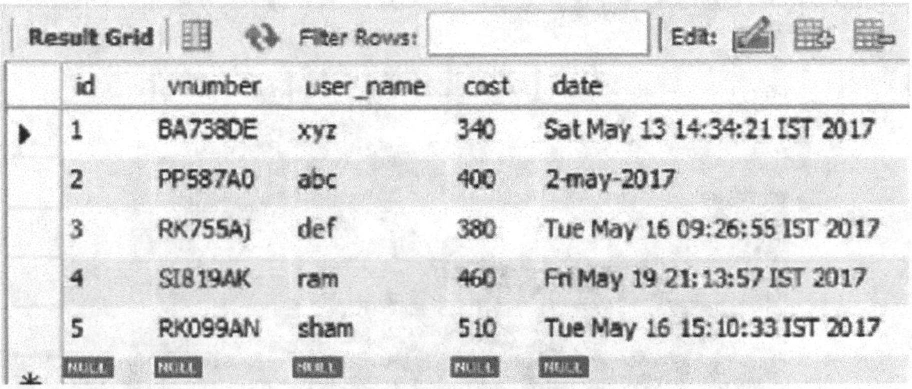

Result Grid | Filter Rows: | Edit:

id	vnumber	user_name	cost	date
1	BA738DE	xyz	340	Sat May 13 14:34:21 IST 2017
2	PP587A0	abc	400	2-may-2017
3	RK755Aj	def	380	Tue May 16 09:26:55 IST 2017
4	SI819AK	ram	460	Fri May 19 21:13:57 IST 2017
5	RK099AN	sham	510	Tue May 16 15:10:33 IST 2017
NULL	NULL	NULL	NULL	NULL

Fig. 6. Toll collection results

5 Conclusion and Future Scope

In this work, existing methodologies and algorithms proposed in literature for Vehicle and Number Plate recognition were reviewed. Due to the unavailability of such an ANPR system off the shelf in tune with our requirements, it is our endeavor to customize an ANPR system for educational institution. OCR was implemented on number plates obtained from static images and an average accuracy of 80% was obtained. This accuracy can be improved greatly by positioning the camera suitably to capture the best frame by using temporal redundancy. The implementation of the proposed system can be extended for the recognition of number plates of multiple vehicles in a single image frame by using multi-level genetic algorithms. Also, a more sophisticated version of this system can be implemented by taking inputs from live video feed and selecting the best vehicle frame for classification of vehicle types and recognizing the number plates using neural networks.

References

1. Kodwani, L., Meher, S.: Automatic license plate recognition in real time videos using visual surveillance techniques. ITSI Trans. Electr. Electron. Eng. TEEE **1**, 60–66 (2013). ISSN (PRINT): 2320-8945
2. Patel, C., Shah, D., Patel, A.: Automatic Number Plate Recognition System (ANPR): a survey. Int. J. Comput. Appl. **69**, 21–33 (2013)
3. Qadri, M.T., Asif, M.: Automatic Number Plate Recognition System for vehicle identification using optical character recognition. In: International Conference on Education Technology and Computer, pp. 335–338. IEEE (2009)
4. Puranic, A., Deepak, K.T., Umadevi, V.: Vehicle number plate recognition system: a literature review and implementation using template matching. Int. J. Comput. Appl. **134**, 12–16 (2016)
5. Ansari, N.N., Singh, A.K.: License number plate recognition using template matching. Int. J. Comput. Trends Technol. **35**, 175–178 (2016)
6. Han, B.G., Lee, J.T., Lim, K.T., Chung, Y.: Real-time license plate detection in high resolution videos using fastest available cascade classifier and core patterns. ETRI J. **37**, 251–261 (2015)

A Co-design Methodology to Design and Develop Serious Educational Games: Integrating the Actors and Challenges in Each Phase

Neha Tuli[1(✉)] and Archana Mantri[2]

[1] Department of Computer Science and Engineering, Chitkara University Institute of Engineering and Technology, Chitkara University, Rajpura, Punjab, India
neha.tuli@chitkara.edu.in

[2] Department of Electronics and Communication Engineering, Chitkara University Institute of Engineering and Technology, Chitkara University, Rajpura, Punjab, India
archana.mantri@chitkara.edu.in

Abstract. The use of Serious Games (SGs) for education is the latest trend and an important opportunity to improve education, providing the students with realistic and immersive simulations of real-time environments. This paper presents an enhanced model for design and development of serious educational games that integrates pedagogical and game design aspects together along with the role distribution of the actors (e.g. teachers, game designers, developers, students) involved in the development process. Development of educational games not only requires extensive knowledge of the learning objectives but also the ability to make the game fun and engaging activity for the students. This paper also highlights some of the important issues faced while development which needs to be analyzed to support the research in SG development. We believe that our research work would help other researchers and game developers to understand the design process and the related issues to develop better serious educational games to aid teaching in schools.

Keywords: Serious games · Serious educational games · Serious game design
Interactive learning environments
Issues and challenges in serious game development

1 Introduction

With the technology advancements, the games are now being used in various fields, the biggest being education, giving positive outcomes by improving teaching-learning processes. Serious educational games have a unique ability to frame an understanding of the concepts by student participation and immersion rather than just observation. Serious Games are the games that are designed and developed for some purpose other than just entertainment. Educators and Researchers are trying to merge SGs into school

R. Sharma et al. (Eds.): ICAN 2017, CCIS 805, pp. 42–52, 2018.
https://doi.org/10.1007/978-981-13-0755-3_4

since their emergence [1]. It is assumed that immersive and interactive features of games with problem-solving approach have the ability to enhance learning and teaching [2].

2 Background

Games have become a significant teaching tool as they offer a convincing environment via interactive, engaging and immersive activities. The use of SGs in education can have a very positive influence providing engaging learning environments [3]. Barbosa and Silva described a SG to teach basic functionality of human circulatory system [4]. Muratet et al. presented a game to help students learn programming concepts [5]. Garris et al. suggested that there are certain games, which have empirically demonstrated that they can be effectively used in classrooms for teaching [6]. Despite the advancements in serious educational games, many reports [7] have shown that their development is not an easy task. Peery discussed issues and constructs that the SGs developers must be aware of while making a successful educational game. They mainly discussed how learning gains should be associated with the game in accordance to bloom's taxonomy [8]. Peng described some issues and challenges while developing a game including students concerns and selection of game engines for developing games. He also compared different game development tools based on game type and features, programming required and the development environment and favored Unity 3d game engine [9]. Barbosa et al. proposed a new methodology of design and development of SGs that integrates educational content in the games [10]. The methodology discussed by them proposes to integrate the learning mechanisms (e.g. quiz/puzzles/mini-games etc.) to each level of the game so that the students learn while completing such tasks. They did not discussed the role of actors and the challenges experienced in each phase of developing SGs. Very less is known about methods, concepts and tools that are presently used in creating serious games. The challenges and issues faced in the development process and the methods to deal with these challenges are yet to be explored. In this paper, we propose a new methodology of developing SGs, which integrate the roles of actors and the challenges faced in each design/development process.

3 Design and Development Process

The game development process of entertainment games includes Concept Development, Designing, Implementation, Testing and Deployment (Fig. 1).

Concept Development includes storyboarding of the theme/concept of the game. It describes what the whole game is about and how it has to be developed. Designing includes designing of User Interface, 3D models, animations etc. required for the project. Implementation includes actual development of codes and algorithms to make things work together. Testing and Deployment include analyzing the game to identify any bugs or errors, after which the final build of the game is deployed.

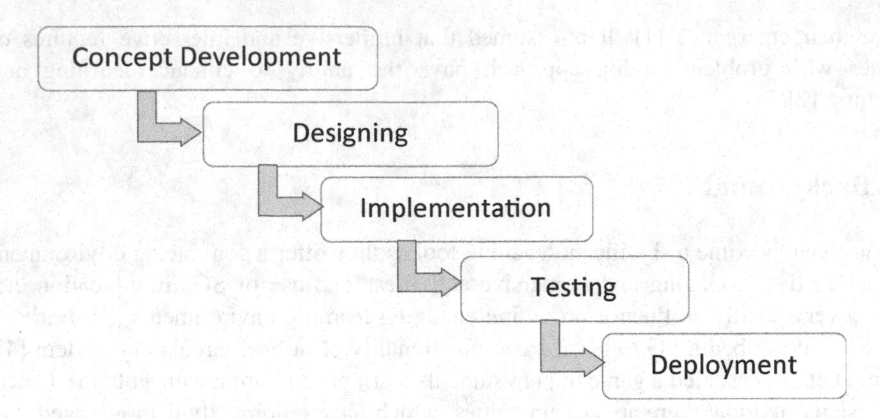

Fig. 1. Design and development process of entertainment games

The development of entertainment games is completely left on the developer's understanding. Only story boarders and the developers can handle the development process. There is no need of any domain/subject expertise for assistance in the whole process. Whereas, Serious Educational Game development requires input from actors (domain experts, story boarders, designers, developers). The main motive of developing SEGs is providing education with entertainment. The game design and development process in serious games requires inherent involvement of educators as opposed to non-serious/entertainment games. Marne et al. discussed that SGs depend on the developer's efficiency and the educator's competence and expertise. The major problem in developing an effective SG, which has both the learning as well as entertainment, is to create a common setting for the teacher and designer to communicate [11]. It is a challenging task for the designer and the developer to think like a teacher to develop an effective and usable game for the students. Therefore, we propose an enhanced SEG development process (Fig. 2).

The proposed method defines all the phases of the game development process in accordance to the serious educational games. The proposed model also highlights the challenges and issues that are faced during each phase of the development process. These challenges must be considered while developing an educational game to make the development process easier.

3.1 Phase 1: Concept Development/Storyboarding

Storyboarding includes all the details of how the game is to be developed including the concept, scope and value of the game. To make an effective SEG, certain steps are to be designed to associate the learning objectives of the topic with the games. Three steps are proposed for this research work.

The **first step**, involves answering certain questions that must be considered while associating learning gains with the game. The questions include –

- what are the learning gains of game,
- how to associate pedagogical content with game activities,

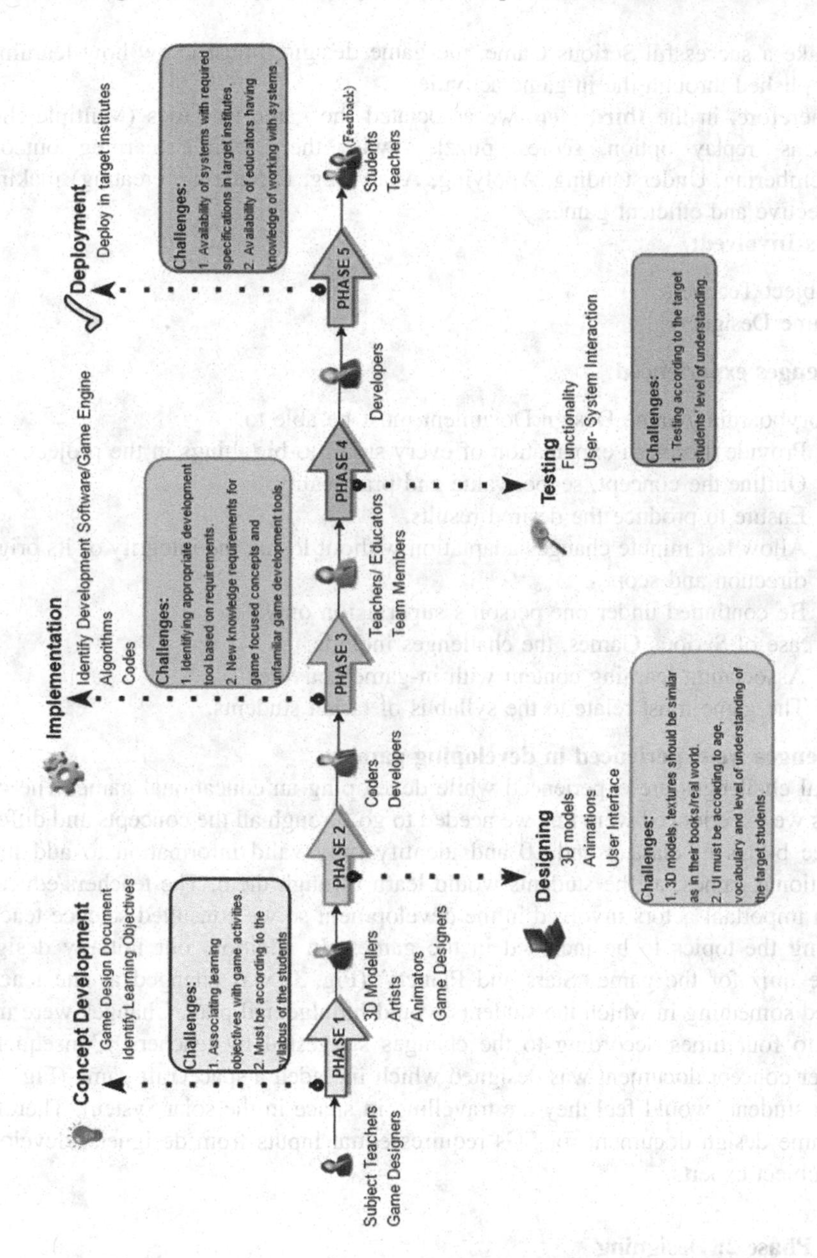

Fig. 2. Proposed method of design and development process of SEGs

- who are the actors involved in the development process,
- what roles would the educators play in the deployment and
- on what metrics the efficiency of the game would be measured.

The **second step** includes the listing of the game attributes of the game by identifying the category, rules, elements, motivations and tasks associated with the game.

To make a successful Serious Game, the game designer must know how learning is accomplished through the in-game activities.

Therefore, in the **third step**, we associated the game activities (Multiple-choice questions, replay option, scores, puzzles) with the Bloom's learning outcomes (Remembering, Understanding, Applying, Analyzing, Evaluating, creating) making it an effective and efficient game.

Actors Involved:

- Subject Teachers
- Game Designers

Challenges experienced:

- Storyboarding/Game Design Document must be able to:
  - Provide thorough explanation of every small to big things in the project.
  - Outline the concept, scope, value and practicality.
  - Ensure to produce the desired results.
  - Allow last minute changes adaptation without losing the integrity of its original direction and scope.
  - Be continued under one person's supervision only.
- In case of Serious Games, the challenges include:
  - Associating learning content with in-game activities
  - The game must relate to the syllabus of target students.

Challenges we experienced in developing games:

Several challenges are experienced while developing an educational game. The challenges we experienced were that we needed to go through all the concepts and different science books of class 8 and 10 and identify only valid information to add in the educational games, as the students would learn through them. The teachers/educators are an important actors involved in the development so we consulted science teachers teaching the topics to be included in the games. In addition, our initially designed simple quiz for the game 'Stars and Planets' (Fig. 3) was changed as the teachers wanted something in which the students would indulge and play. Changes were made three to four times according to the changes suggested by teachers. Consequently, another concept document was designed which included a spacecraft game (Fig. 4) in which students would feel they are travelling in space in the solar system. Therefore, the game design document for SGs requires equal inputs from designers, developers and subject experts.

3.2 Phase 2: Designing

This phase includes designing models, animations, User Interface used in the game which were developed using software Autodesk Maya. UI refers to the approaches (keyboard control, mouse control) and interfaces (inventory screen, map screen) through which a user interacts with your game. Some of the factors influencing the acceptance of game based learning include Usability and Interactivity [12]. Developers/Programmers do not take up this job as they feel it to be a creative work and is mostly done by the game designers (modelers and animators). Interface appeal is

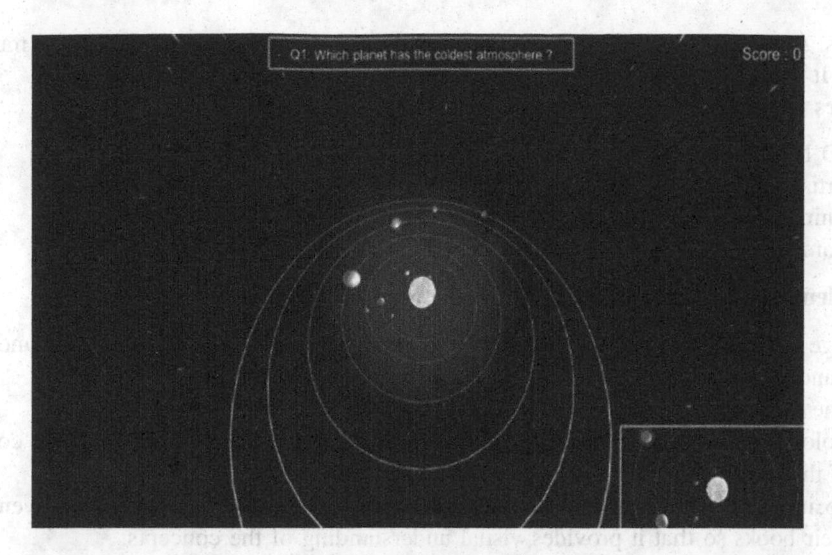

Fig. 3. Before changes

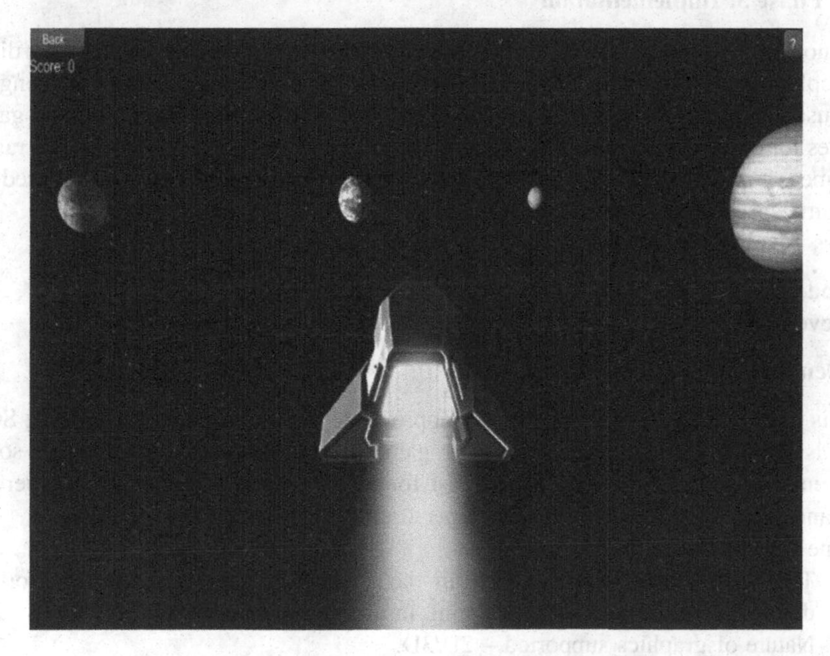

Fig. 4. After changes

the important aspect and it must be engaging to play having graphical attractiveness, appropriate/relevant instructions displayed in place and time according to their requirements and should focus on certain functional aspects like how big is the screen, what data is shown and where, how the player navigates through it and use two or three

colors. A game may lose its importance by providing very less information or too much of it. It must be easy for a new player to interact with the game elements.
Actors Involved:

- 3D Modelers
- Artists
- Animators
- Game Designers

Challenges experienced: We need to take care of:

- Age of students for whom we are developing the game and their level of understanding (which may differ from one student to another).
- The vocabulary they are able to understand.
- Color combinations that would help them recall afterwards (e.g. we kept the color of the planet and their description font color same).
- Positions of the windows, same textures of planets and lab equipment as given in their books so that it provides visual understanding of the concepts.

3.3 Phase 3: Implementation

It is another important phase of development process. After studying the storyboarding, the implementation of the games begin with identifying the software and game engine to be used in the design and development of SEGs. There are number of tools and game engines for creating games. Although there are various tools for developing wide range of games, it is very challenging for the developer to make the best selection based on the game specifics and mechanics [13].
Actors Involved:

- Coders
- Developers

Challenges experienced:

- It is a challenging task to identify the appropriate game engine for developing SGs.
- It is difficult to say that a particular game engine is perfect as all have some strengths and weakness. Some are fast for development, some may have performance issues, and some have developer friendly interface.
- One must identify:
  - The set of target platforms to run the game that requires minimal effort in development process and integrating in different platform.
  - Nature of graphics supported – 2D/3D.
  - Easy testing.
  - Low cost.
  - User Interface – easy or difficult to understand.
  - Availability of documentation.
  - Previously developed games.

3.4 Phase 4 and Phase 5: Testing and Deployment

After development, the team members, educators, test the game. After making the suggested changes, it is deployed in target schools/institutes for the students to learn and play.

Actors Involved (Testing):

- Subject Teachers/Educators
- Team Members

Actors Involved (Deployment):

- Developers

Challenges experienced:

- Availability of systems, software, and hardware as required for the efficient working of the games.
- Availability of educators having desired knowledge of working with different software.

4 Results

Using the proposed game development process, we developed two serious games for class 8th (Stars and the Planets) and 10th (Sources of Energy). The "Stars and the Planets" game aims to teach about solar system, seasons and phases of moon in the form of in-game puzzles and activities (Fig. 5). The "Sources of Energy" game provides knowledge about thermal energy, wind energy by giving students virtual lab environments to perform experiments related to topics (Fig. 6).

Fig. 5. Screenshot of Stars and the Planets game

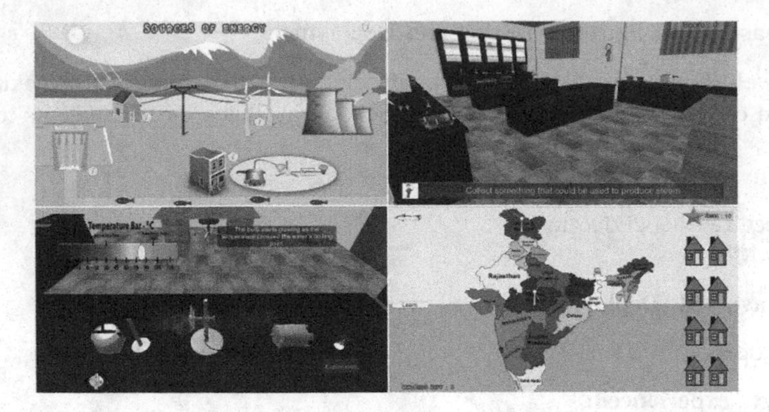

Fig. 6. Screenshot of Sources of Energy game

The game was given to 30 students to play. After deployment, we took feedback from students in the form of questionnaire (Fig. 7) through which we could get any suggestions regarding the improvement if the game, so that we can improve the game and deploy it on larger audience. It was found that the students liked the games and enjoyed playing it repeatedly as the learning content was well integrated in the game.

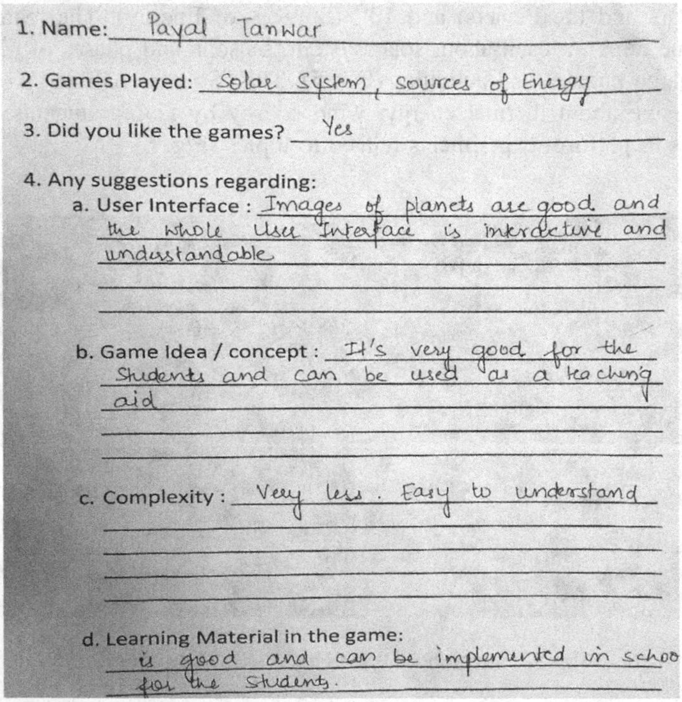

1. Name: Payal Tanwar

2. Games Played: Solar System, Sources of Energy

3. Did you like the games? Yes

4. Any suggestions regarding:

a. User Interface : Images of planets are good and the whole User Interface is interactive and understandable

b. Game Idea / concept : It's very good for the Students and can be used as a teaching aid

c. Complexity : Very less. Easy to understand

d. Learning Material in the game: is good and can be implemented in schoo for the Students.

Fig. 7. Feedback from Student-1

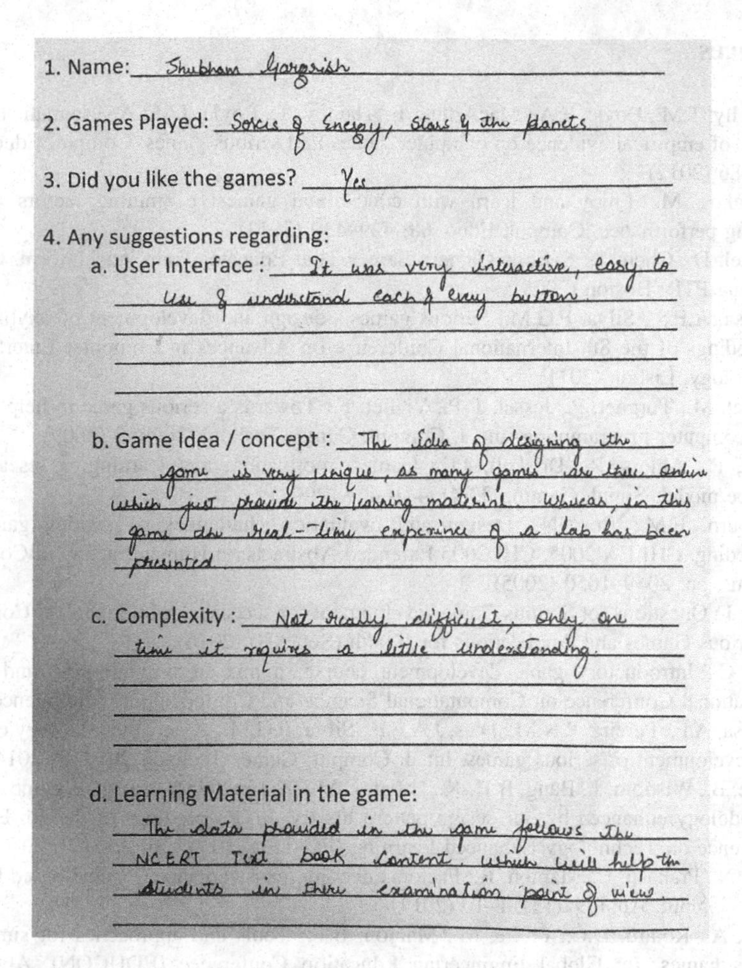

1. Name: Shubham Gargrish

2. Games Played: Sources of Energy, Stars & the Planets

3. Did you like the games? Yes

4. Any suggestions regarding:

a. User Interface : It was very interactive, easy to use & understand each & every button.

b. Game Idea / concept : The idea of designing the game is very unique, as many games are there online which just provide the learning material. Whereas, in this game the real-time experince of a lab has been presented.

c. Complexity : Not really difficult, only one time it requires a little understanding.

d. Learning Material in the game: The data provided in the game follows the NCERT Text book content which will help the students in their examination point of view.

Fig. 8. Feedback from Student-2

Some of the students suggested us to add more music to the game. In addition, we observed that students liked the UI of the game and they were able to understand the flow of the game and instructions very easily (Fig. 8).

5 Conclusion and Future Scope

In conclusion, our games proved to be an effective supplement for classroom teaching. Still, we deployed the game on limited audience to test its efficacy and know its limitations in terms of learning content or User Interface. The deployment results showed that the game proved to be useful for the teachers helping them in their lectures. In future, we would test our game in broader audience and calculate its effectiveness in more detail.

References

1. Connolly, T.M., Boyle, E.A., MacArthur, E., Hainey, T., Boyle, J.M.: A systematic literature review of empirical evidence on computer games and serious games. Comput. Educ. **59**(2), 661–686 (2012)
2. Giannakos, M.: Enjoy and learn with educational games: examining factors affecting learning performance. Comput. Educ. **68**, 429–439 (2013)
3. Michael, D., Chen, S.: Serious Games: Games That Educate, Train, and Inform. Cengage Learning PTR, Boston (2005)
4. Barbosa, A.F.S., Silva, F.G.M.: Serious games—design and development of oxyblood. In: Proceedings of the 8th International Conference on Advances in Computer Entertainment Technology, Lisbon (2011)
5. Muratet, M., Torguet, P., Jessel, J.-P., Viallet, F.: Towards a serious game to help students learn computer programming. Int. J. Comput. Games Technol. **2009**, 3 (2009)
6. Garris, R., Ahlers, R., Driskell, J.E.: Games, motivation, and learning: a research and practice model. Simul. Gaming **33**(4), 441–467 (2002)
7. Raybourn, E.M., Bos, N.: Design and evaluation challenges of serious games. In: Proceeding, CHI EA 2005, CHI 2005 Extended Abstracts on Human Factors in Computing Systems, pp. 2049–1050 (2005)
8. Peery, J.: Questions for Serious Game development for success. In: International Conference on Serious Games and Applications for Health (SeGAH) (2016)
9. Peng, C.: Introductory game development course: a mix of programming and art. In: International Conference on Computational Science and Computational Intelligence (2015)
10. Barbosa, A.F., Pereira, P.N.M., Dias, J.A.F.F., Silva, F.G.M.: A new methodology of design and development of serious games. Int. J. Comput. Games Technol. **2014**, 8 (2014)
11. Marne, B., Wisdom, J., Bang, B.H.-K., Labat, J.-M.: The six facets of serious game design: a methodology enhanced by our design pattern library. In: Proceedings of the 7th European conference on Technology Enhanced Learning (EC-TEL 2012) (2012)
12. Saleh, N., Prakash, E., Manton, R.: Factors affecting the acceptance of game-based learning. Int. J. Comput. Appl. **92**(13), 1–10 (2014)
13. Calvo, A., Rotaru, D.C., Freire, M., Manjon, B.F.: Tools and approaches for simplifying serious games. In: Global Engineering Education Conference (EDUCON), Abu Dhabi (2016)

VLSI Floorplanning Using Entropy Based Intelligent Genetic Algorithm

Amarbir Singh[1(✉)] and Leena Jain[2]

[1] I. K. Gujral Punjab Technical University, Jalandhar, India
singh.amar.gill@gmail.com

[2] Global Institute of Management and Emerging Technologies, Amritsar, India
leenajain79@gmail.com

Abstract. Very Large-Scale Integrated (VLSI) floorplanning is NP-hard combinatorial optimization problem and it is vital in chip design as it determines the quality of chips. To solve this problem in an effective manner, an intelligent approach based on heuristic placement strategy and entropy based genetic algorithm is proposed in this paper called Entropy Based Intelligent Genetic Algorithm (EBIGA). In the proposed work, concept of entropy is introduced in genetic algorithm in order to resolve the problem of local optimal solution. An integer coding representation is used in this paper which makes the task of representation of modules simple. The experimental results on Microelectronics Centre of North Carolina (MCNC) and Gigascale Systems Research Centre (GSRC) benchmarks demonstrate that EBIGA can achieve the optimal and competitive solutions for both fixed-outline and outline-free floorplans.

Keywords: VLSI floorplanning · Entropy · Genetic algorithm

1 Introduction

In the Very Large Scale Integrated (VLSI) circuit design process, floorplanning is an important stage. After partitioning stage, the complex circuit is partitioned into smaller circuits in the next stage called Floorplanning. It is done in order to decide relative locations of the different modules within each sub-circuit with the objectives of minimizing the total wasted space (dead space) in the layout or minimizing the total chip area used by a circuit and minimizing the interconnecting wire length between modules. Various stages in physical design of ICs are Partitioning, Floorplanning, Placement, Routing and Compaction.

A floorplan is a rectangular dissection which describes the relative placement of electronic modules on the chip. The development of integration technology has followed the famous Moore's Law [1]. Gordon Moore, the co-founder of Intel, in the year 1965, stated that "The number of transistors per chip would grow exponentially (double every 18 months)". In the design of VLSI circuits, floorplanning helps to determine the topology of layout and this problem is known to be NP-hard (Non-Deterministic Polynomial-time Hard). Due to the usefulness and complexity, this problem has received much attention in recent years. The solution space of this problem increases exponentially with the increase in number of modules in a circuit. Due to this it

R. Sharma et al. (Eds.): ICAN 2017, CCIS 805, pp. 53–71, 2018.
https://doi.org/10.1007/978-981-13-0755-3_5

becomes very difficult to find the optimal solution by exploring the global solution space [2]. In the past, numbers of metaheuristic algorithms have been used to solve this problem. In this paper Entropy Based Intelligent Genetic Algorithm (EBIGA) is proposed to solve the VLSI floorplanning problem for non-slicing floorplans and hard modules.

In the proposed algorithm, concept of entropy is used in genetic algorithm to measure the diversity of population. The concept of entropy was given by one of the electrical engineer claude Shannon and a recent research in this field reveals that it can be used in different areas like physics, biology, economics, computer science and engineering etc. The entropy represents a measure of uncertainty in random variable or random function. In other words, entropy represents a function which endeavors to illustrate its unpredictability. Entropy is maximum when the distribution of random variable is equal and it becomes minimum, when each random variable has unequal probability. Entropy of random variable Z with a probability distribution P(Z) is maximum, when all the probabilities are same, it can be represented mathematically as below [3].

$$H(P) = -K\sum\nolimits_{i=1}^{n} p_i log p_i \tag{1}$$

Where K is a positive constant.

Recently, various stochastic optimization algorithms, such as Simulated Annealing (SA), Genetic Algorithm (GA), artificial neural networks and tabu search have been used to solve complex problems. It has been proved that GA is an effective method for solving NP-hard optimization problems, so due to this reason GA has been applied on VLSI Floorplanning problem in our research. We have also used the concept of information entropy as a measure of diversity of each population into the process of GA to solve the problem of falling into a local optimal solution. Due to the constant increase in design complexity of the VLSI circuits, floorplanning phase needs to handle more constraints such as fixed-outline and alignment of individual modules. These constraints are also taken into consideration in the proposed solution.

Rest of the paper is organized as follows. Preliminaries to VLSI floorplanning problem are discussed in Sect. 2. In Sect. 3, problem statement is given along with the various constraints. Heuristic Placement strategy and EBIGA is discussed in detail in Sect. 4. Experimental results for outline free floorplanning and fixed outline floorplanning on benchmark problems are presented in Sect. 5. Finally, some concluding remarks are given in Sect. 6.

2 Preliminaries

2.1 Floorplan Representations

Floorplans of VLSI circuits can be categorized into two groups, slicing floorplans [6, 7] and non-slicing floorplans [4, 5] as shown in Figs. 1 and 2 respectively. Any floorplan is sliceable if its rectangular dissection can be obtained by recursively dividing rectangles into smaller rectangles until each non-overlapping rectangle is invisible. So such floorplans which are not sliceable are called non sliceable floorplans. A binary tree representation can be used for a floorplan with slicing structure [8]. The slicing floorplan is less useful than non-slicing floorplan, as it can be used to solve the floorplanning problems with rather limited scope [9]. Non slicing floorplan is much more suitable in VLSI floorplanning and it can generate any type of layout in the practice applications, however because of its non-slicing structure, it cannot be modeled using a slicing tree or binary tree [10]. Due to this, many researchers pay attention to the non-slicing floorplan representations. Different representations of non-slicing floorplan have been proposed in recent years: Corner Block List (CBL) representation [11]; Sequence Pair (SP) representation [12]; Bounded Slicing Grid (BSG) representation [13]; O-tree representation [14]; Transitive Closure Graph (TCG) representation [15], B* tree representation [16, 17] etc. Literature [18] introduces integer coding representation, which is simple, effective and quite suitable for floorplanning problem, due to these advantages the integer coding representation has been adopted in this research work. Further details of the above mentioned representations can be found in literature [10].

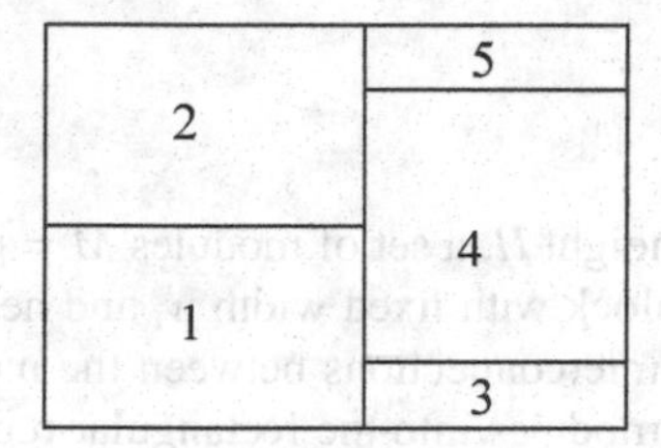

Fig. 1. Slicing floorplan

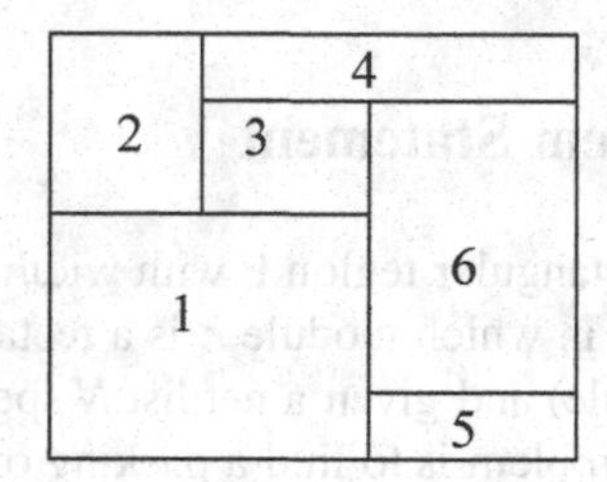

Fig. 2. Non-Slicing floorplan

2.2 Floorplan Constraints

In the classical outline-free floorplanning, the prime objective is to minimize the total floor area without any fixed-outline constraints. Due to the advancement in technology, the modern VLSI design demands fixed die (Fixed-Outline) floorplans [19], so the floorplan optimization problem needs to be formulated with fixed-outline constraints. Based on the sequence pair representation [12], Adya and Markov [20, 21] presented new approach for the floorplan problem [22] to better guide the local search for fixed-outline floorplanning. In the proposed work, fixed-outline constraints are also taken into consideration for designing the floorplan.

2.3 Various Optimization Techniques

Recently lots of researchers have resorted to efficient iterative optimization algorithms in order to search for an optimal floorplan. An iterative algorithm is run until a feasible floorplan is achieved or no more improvements can be obtained. The most popular iterative optimization algorithms used to solve VLSI floorplanning problem are Genetic

Algorithm (GA) [18, 23–26], Simulated Annealing (SA) [17, 27–30] and Particle Swarm Optimization (PSO) [2, 9, 31]. In this research paper, GA is used to find the optimal solution but one of the problems with genetic approach is local optimal solution. This problem may occur when the whole population of a generation becomes similar and it leads to low diversity population. To resolve the problem of local optimal solution Renyi's entropy is used in this paper to ensure that crossover and mutation operators of genetic algorithm generate a generation with different population. Mathematically Renyi's entropy over a probability distribution P = P_1, P2, P_3, …, P_n is defined as

$$R(P) = \frac{1}{1-\alpha} l_n \sum\nolimits_{n=1}^{k} p^{\alpha} \quad (2)$$

Where $\alpha > 0$ and $\alpha \neq 1$.

So with the help of Renyi's entropy a new algorithm called Entropy Based Intelligent Genetic Algorithm (EBIGA) is devised to find a near to-optimal solution to the VLSI floorplanning problem using the both outline free and fixed-outline constraints with the primary objective of minimizing the dead space.

3 Problem Statement

Given a rectangular region P with width *W* and height *H*, a set of modules $M = \{r_1, r_2, r_3, \ldots, r_m\}$, in which module r_i is a rectangular block with fixed width w_i and height h_i (hard module) and given a net list *N* specifying interconnections between the modules in *M*, the problem is to find a packing of all the modules into the rectangular region P, such that they meet the following conditions:

(1) No module can be aligned in a diagonal position; it must be placed parallel to the coordinate axis.
(2) Overlapping of one module with any other module is not allowed.
(3) All modules must lie inside the rectangular region P.
(4) A module may be rotated by 90° before its placement.

Given *M* and *N*, goal of the floorplanning problem is to find a floorplan *F* such that a cost function is minimized [30]. The cost of a floorplan *F*, *cost* (*F*), consists of two parts, one is the area, *area* (*F*), which is measured by the smallest rectangle that encloses the floorplan and the other is *wirelength* (*F*), which is the wirelength of all the nets specified by *N or* the interconnection cost between various modules. The wirelength of a net is calculated by the half perimeter of the minimal rectangle in the floorplan, which encloses the centers of the modules connected by the net. Now the cost function can be defined as follows:

$$cost(F) = (w) \times \frac{area(F)}{norm\_area} + (1-w) \times \frac{wirelength(F)}{norm\_wire} \quad (3)$$

Where w is a weight assigned for primary objective related to area, $w \in [0, 1]$, *norm_area* and *norm_wire* are the minimal area and the minimal wirelength cost of the problem respectively. This kind of cost function has been adopted from [23]. Since we do not know *norm_area* and *norm_wire* in practice, estimated values are used.

3.1 Floorplanning Under Fixed Outline Constraints for Hard Modules

The fixed-outline constraint for floorplanning problem is implemented in the following fashion. For a given fixed-outline, the desired aspect ratio (α), that is, height/width is fixed.

For a solution y the percentage of dead space (Γ) is computed as below;

$$\Gamma = \left[\frac{\mathit{Floorplan\ Area} - \mathit{Sum\ of\ module\ area}(A)}{\mathit{Floorplan\ Area}}\right] \times 100 \tag{4}$$

The values of height H^* and width W^* of outline are defined in the literature [21] and are given below;

$$H^* = \sqrt{(1+\Gamma)A\alpha} \tag{5}$$

$$W^* = \sqrt{(1+\Gamma)A/\alpha} \tag{6}$$

The objective function defined in Eq. (3) is subjected to the constraint;

$$H \leq H^* \text{ and } W \leq W^* \tag{7}$$

Where W and H are the width and height of any feasible floorplan y and solutions satisfying the fixed-outline constraint (7) are considered as feasible.

3.2 Integer Coding Representation

Literature [18] introduces integer coding representation in the format of $<V_1, V_2, \ldots, V_i, \ldots, V_n>$, where $1 \leq V \leq n$, $V = j$ and it denotes that the ith module is place at j^{th} positon. Then a heuristic operator is needed to adjust the modules while taking both their shapes and alignment into consideration. Literature [32] adopted the integer coding representation proposed in [18] and modified the heuristic adjustment method. In this paper, the representation is similar to that in [18], but it only contains the order of modeling, and it does not contain the information of shape and alignment of the modules.

3.3 Basics of Genetic Algorithm

A genetic algorithm is a stochastic optimization search approach proposed by Holland in 1962 and it is based on the survival of the fittest principle. It mimics the natural selection and evolutionary genetic mechanisms of human beings and transfers the best genes from one generation to another. In every generation selection, crossover and mutation operators are applied to produce a new offspring having some of the characteristics of their ancestors required for survival. In this way a genetic algorithm can help to find an optimal solution when the solution space is too large. The fitness of a floorplan (Individual) F is defined as follows in our work.

$$fitness(F) = \frac{1}{cost(F)} \tag{8}$$

Where cost (F) is the cost of floorplan F defined in (3).

4 Heuristic Placement Strategy and Optimization Using EBIGA

In this section, the overall procedure of placement of modules and the approach used to optimize the floorplan is discussed.

4.1 A Novel Heuristic Placement Strategy

The module sequences for placement will be generated during the creation of population for genetic algorithm, and some rules are required to convert modules sequence into dimensional floorplan. Each module sequence will be converted into a floorplan according to the rules given below:

- Place the first module at the left bottom corner according to the sequence of modules generated by genetic algorithm. Two pivot points are identified on the introduction of every module.
- For each valid pivot point, the next module in sequence is placed on the chip in both length and Width wise orientation and area of the resulting rectangular casing is noted. The pivot point, which gives the smallest enclosing rectangle, is selected.
- The algorithm needs to check that the newly placed module does not overlap with the boundary of the rectangular chip or with the modules placed earlier before the floorplan is accepted as valid.
- The latter step is repeated for all the remaining modules in the sequence.

4.2 Modified One Point Ordered Crossover Operator (MOOC)

Inspired by the GA, a new crossover operator for the floorplan problem is proposed in this paper called Modified One Point Ordered Crossover Operator (MOOC). This crossover operator is influenced from one point crossover operator and ordered crossover operator. Suppose there are two parent strings: $(5, 3, 10, 9, 8, 6, 4, 7, 2, 1)$, and $(6, 9, 2, 7, 1, 5, 4, 8, 10, 3)$ and cut point lies between third and fourth bit, then the offspring are created in a following way. Starting from the beginning of one parent up to cut point values are copied into the offspring, which give $(5,3,10,*,*,*,*,*,*,*)$, then rest as of the values are copied in the order in which they appear in other parent and omitting the values that are already present. It give us the first offspring as $(5,3,10,6,9,2,7,1,4,8)$. For the second offspring, starting from the cut point values are copied from other parent into the offspring while preserving the initial position of values, which give $(*,*,*,7,1,5,4,8,10,3)$. Then rest of the values is copied in the order in which they appear in other parent and omitting the values that are already present. It give us the second offspring as $(9,6,2,7,1,5,4,8,10,3)$. This crossover operator helps to preserve the head portion of one parent and tail portion of other parent as these parts generally have important information in relevance to the optimal solution of the problem.

4.3 Entropy Based Intelligent Genetic Algorithm

Measure of Diversity of a Chromosome Population. Genetic algorithm does not guarantee to find an optimal solution, but the past experience shows that genetic evolution based approach is very useful in solving VLSI problems in an efficient manner [35]. One of the major problems with genetic algorithm is that it may fall into a local optimal solution during the evolution process. Sometimes the count of similar chromosomes starts to rise rapidly in a population and in such a situation mutation operator fails to maintain the diversity of the population. In this paper an Entropy Based Intelligent Genetic Algorithm (EBIGA) is proposed to escape from the trap of above said problem. In EBIGA the diversity of the chromosome population is measured in every generation and the low diversity population is improved with the help of Renyi's entropy. In our algorithm the locus diversity R_i of the ith locus (i = 1, 2, 3, …, n) of a population is measured by using the following formula.

$$R_i = \frac{1}{1-\alpha} l_n \sum\nolimits_{m \in M} pr_{im}^{\alpha} \quad (9)$$

Where $pr_{im} = \frac{na_{im}}{pop\_size}$, $\alpha > 0$ and $\alpha \neq 1$

M: the set of modules that are to be placed

na_{im}: the number of appearance of module m at locus i

The local diversity R_i approaches to the minimum value 0 when the number of appearances of one of the modules is much more than the number of appearances of the other modules. Value of R_i is computed by comparing it with the threshold value.

Complete Procedure of EBIGA.
Step 0: Input Data
Select the MCNC/GSRC benchmark problem to be optimized and set the value of n according to the number of modules in the benchmark data. Set the parameters for GA such as population size *(pop_size)*, crossover probability *(C_p)*, mutation probability *(M_p)*, maximum number of generations *(mgen)* and control parameter (cp).
Step 1: Initial Population
Randomly generate an initial population or chromosomes equivalent to *pop_size*.
Step 2: Calculate the Fitness

Step 2.1: evaluate the fitness value of each chromosome by applying novel Heuristic placement strategy and using *eq. (3)*.
Step 2.2: select the chromosome with largest fitness value and store it as a member of next generation.

Step 3: Crossover
Apply MOOC operation over the chromosomes selected by using C_p.
Step 4: Mutation
Perform the mutation operation on the resulted chromosomes from crossover by using mutation probability *Mp.*
Step 5: Selection
Select chromosomes for the next generation by using tournament selection method.
Step 6: Calculate and improve the Diversity of Population using Renyi's Entropy.

Step 6.1: set *counter* = *0* and $I = \{\phi\}$
Step 6.2: Repeat the following for $i = 1$ *to n.*
Evaluate the locus diversity R_i of each locus of the population by using *eq. (9)*.
If $R_i \leq l_n(\alpha)$, where higher value of α can increase the cardinality of a set *I*.
Then add the index *i* to the set *I* and increment *counter*.
Step 6.3: if $counter \geq \frac{n}{cp}$ then go to Step 6.4, else go to Step 7. Here value of *cp* can be any integer value in the closed interval [2, n] and higher value of *cp* means higher probability of doing improvement.
Step 6.4: Generate a random integer number *rnd* in the range $[\frac{pop\_size}{cp}, pop\_size - 1]$ and extract *rnd* chromosomes randomly from the population. In each of the chromosomes selected, exchange genes randomly on the positions included in set *I* so that diversity in selected population is increased and go to Step 6.1.

Step 7: Stop checking
If *mgen* generations have been processed, then print output and stop. Otherwise go to Step 2.

5 Experimental Results

In this part, two experiments are carried out one for classical (outline-free) Floorplanning and another for floorplanning under fixed-outline constraints. MCNC and GSRC benchmarks have been used to test and compare the results of experiments. Characteristics of MCNC and GSRC benchmark circuits are shown in Tables 1 and 2 respectively. The cells of both problems are set as hard modules. MATLAB 7.12 is used to implement and run the programs on a machine with an intel CPU running at 1.70 GHz and 4 GB RAM. In both experiments, the primary objective is to minimize the wastage area (deadspace). The fitness function is described in Eq. (3) and the parameters of the EBIGA are set as follows: The value of 'w' in Eq. (3) is set to 1. Population size for genetic algorithm varies for different benchmark circuits, crossover probability *(C_p)* are set as 0.8 and mutation probability *(M_p)* is 0.1. In the implementation of EBIGA, in order to calculate renyi entropy the value of α is taken as 7 and ln(α) i.e. ln(7) = 1.946 is used as the threshold value. The value of control parameter (cp) is different for every benchmark circuit, as the number of modules is different in each benchmark circuit. Higher value of cp means higher probability of doing improvement. In case of Xerox benchmark circuit, the value of Xerox benchmark circuit is taken as 5.

Table 1. The Chracteristics of MCNC benchmark circuits

Benchmark	Circuit	#Cells	#Nets	Cell area (mm^2)
MCNC	Apte	9	97	46.56
	Xerox	10	203	19.35
	Hp	11	83	8.83
	ami33	33	123	1.156
	ami49	49	408	35.445

Table 2. The Chracteristics of GSRC benchmark circuits

Benchmark	Circuit	#Cells	Cell area (mm^2)
GSRC	n30	30	0.2085
	n50	50	0.1985
	n100	100	0.1795

5.1 Outline-Free Floorplanning with Minimizing Area for Hard Modules

In order to show the efficiency of the algorithm for Classical (outline-free) floorplanning with the objective of minimizing area, we have compared the results with Pang et al. (2000) [33], Chang et al. (2000) [16], Lin et al. (2001) [15], Hong et al. (2001) [11], Tang et al. (2001) [34], Adya et al. (2003) [20] and Chen et al. (2007) [35]. Results of these floorplanning algorithms are taken from their published reports in literature. The wastage or dead-space of a floorplan layout is obtained by calculating the difference between the minimum area of the rectangle which covers all the modules and the sum of all the modules areas. Relative wastage is defined as the ratio of wastage to the minimum area of the rectangle which encapsulates all the modules. Normalization of results is used to compare our results with other benchmark algorithms in terms of area. The normalization is a ratio of the results of other placement algorithms to the results of our algorithm. Some abbreviations are defined for convenience, such as NRW for normalized relative wastage and NRT for normalized relative run time. The simulation results for area are shown in Table 3. The proposed algorithm (EBIGA) produces very competitive results in comparison with other algorithms proposed in literature by different researchers for area minimization. For problems with small number of modules such as apte and Xerox benchmark, the proposed method can achieve better result than Adya et al. (2000) and Hong et al. (2001). However, when the problem scale becomes larger, EBIGA has a better performance than Pang et al. (2000) and Hong et al. (2001) for ami33 and ami49 benchmark respectively. Results for runtimes of various algorithms are provided in Table 4. If time constraint is also taken into consideration along with the area then our algorithm can be a better choice than most of the other algorithms for apte, Xerox and hp benchmarks. Figure 3 shows the percentage of better, same and worse runtimes for EBIGA. Figure 4 show the simulation results of MCNC ami33 benchmark and Fig. 5 show the simulation results of GSRC n30 benchmark. Table 5 shows the values of Renyi entropy (R_i) for xerox benchmark during the first iteration of GA. From the results of renyi entropy it is clear that column number 4, 6 and 8 have lesser renyi entropy values than the threshold value ln(7) = 1.946, so there is a need to improve diversity in these columns for randomly selected set of our population and this process will be repeated until desired level of diversity in population is not reached.

Table 3. Classical (outline-free) floorplanning with the objective of minimizing area

Circuit →	Apte		Xerox		hp		ami33		ami49	
Algorithm ↓	Area (mm^2)	NRW	Area (mm^2)	NRW	Area (mm^2)	NRW	Area (mm^2)	NRW	Area (mm^2)	NRW
EBIGA	46.92	1	20.16	1	9.20	1	1.225	1	37.20	1
DS Chen et al. (2007)	46.92	1	20.26	1.12	9.15	0.86	1.19	0.492	36.30	0.48
SP Parquet (2003) [20]	47.07	1.42	19.83	0.59	9.14	0.84	1.19	0.492	37.27	1.03
Fast SP (2001) [34]	46.92	1	19.80	0.56	8.94	0.30	1.20	0.638	36.50	0.60
CBL (2001) [11]	NA	NA	20.96	1.99	NA	NA	1.20	0.638	38.58	1.79
TCG (2001) [15]	46.92	1	19.83	0.59	8.94	0.30	1.20	0.638	36.77	0.75
B^*-Tree (2000) [16]	46.92	1	19.83	0.59	8.95	0.32	1.27	1.652	36.8	0.77
Enhanced O-Tree (2000) [33]	46.92	1	20.21	1.06	9.16	0.89	1.24	1.217	37.73	1.30

Table 4. Runtimes for various algorithms

Circuit →	Apte		Xerox		hp		ami33		ami49	
Algorithm ↓	Time (sec)	NRT	Time (sec)	NRT	Time (sec)	NRT	Time (sec)	NRT	Time (sec)	NRT
EBIGA	0.24	1	1.51	1	8.01	1	178.9	1	107.84	1
DS Chen et al. (2007) [35]	0.066	0.28	0.1	0.07	0.7	0.09	7.4	0.04	37.03	0.34
SP Parquet (2003) [20]	4	16.67	3	1.99	4	0.50	9	0.05	16	0.15
Fast SP (2001) [34]	1	4.17	14	9.27	6	0.75	20	0.11	31	0.28
CBL (2001) [11]	NA	NA	30	19.87	NA	NA	36	0.20	65	0.60
TCG (2001) [15]	1	4.17	18	11.92	20	2.50	306	1.71	434	4.02
B*-Tree (2000) [16]	7	29.17	25	16.56	55	6.87	3417	19.10	4752	44.07
Enhanced O-Tree (2000) [33]	11	45.83	38	25.16	19	2.37	118	0.66	406	3.76

Table 5. Values of Renyi entropy (R_i) for xerox benchmark during the first iteration of GA.

Module No.	Probability of occurrence of modules at different positions in the population									
	1	2	3	4	5	6	7	8	9	10
1	0.0500	0.1000	0.1500	0.0750	0.0750	0.2000	0.1750	0.0250	0.1250	0.0500
2	0.1500	0.1250	0.1000	0.0500	0.1250	0.0500	0.0000	0.0750	0.1500	0.1500
3	0.1250	0.1500	0.1250	0.0500	0.0500	0.0000	0.1250	0.1750	0.1000	0.1000
4	0.1250	0.1000	0.1000	0.0500	0.1500	0.1000	0.0750	0.0750	0.1250	0.1000
5	0.1500	0.1250	0.1250	0.1250	0.0750	0.0750	0.1000	0.1000	0.0250	0.1250
6	0.0500	0.1000	0.0500	0.0750	0.0750	0.1250	0.1500	0.1750	0.1250	0.1000
7	0.1250	0.1000	0.1000	0.0500	0.1000	0.2000	0.1000	0.0500	0.1000	0.0750
8	0.0750	0.0750	0.0750	0.2000	0.1250	0.0500	0.1250	0.1000	0.0750	0.1000
9	0.0500	0.0750	0.1500	0.1250	0.1000	0.1250	0.0750	0.1500	0.0500	0.0750
10	0.1000	0.0500	0.0250	0.2000	0.1250	0.0750	0.0750	0.0750	0.1250	0.1250
R_i	2.0355	2.1145	2.0451	1.7558	2.0996	1.7553	1.9575	1.8884	2.0787	2.1146

Performance of the proposed algorithm is compared with the other best methodologies reported in the literature such as Simulated Annealing (SA) and Particle Swarm Optimization (PSO). From the Table 6, it can be seen that the results produced by the proposed algorithm for area minimization (w = 1) are better than the results produced by the literature SA [16] and PSO [2] for ami33, n30 and n50 benchmark circuits. For ami49 our results are better than the results produced by an approach proposed in literature [2] using particle swarm optimization. Clearly EBIGA is very competitive and efficient in comparison with some of the best approaches reported in literature.

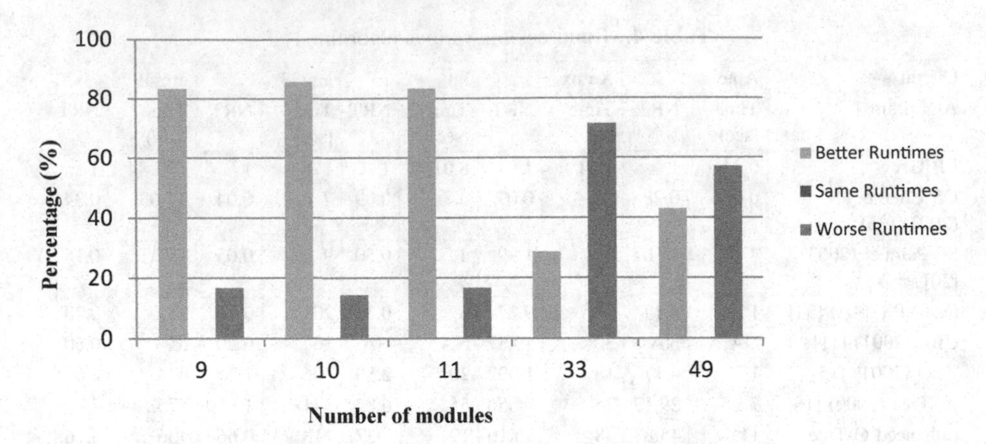

Fig. 3. Percentage of better, same and worse runtimes for EBIGA.

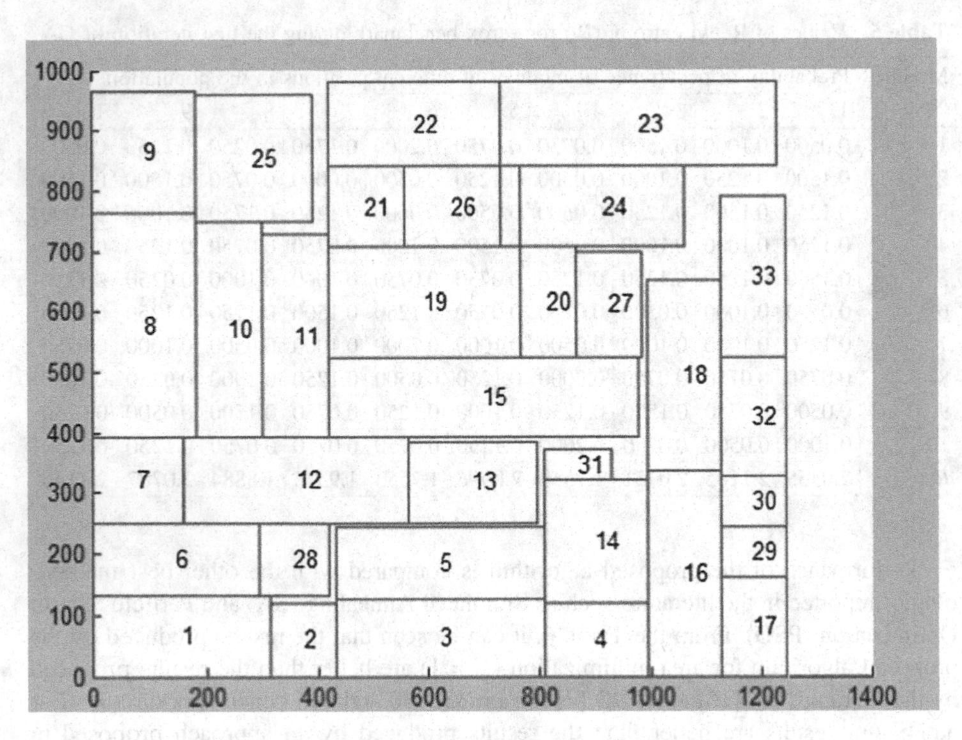

Fig. 4. Simulation result on MCNC ami33

Table 6. Comparison among SA, PSO and EBIGA

Circuit	The number of modules	Area (mm^2)		
		Simulated Annealing [16]	PSO [2]	EBIGA
apte	9	46.92	46.92	46.92
Xerox	10	19.83	20.38	20.16
Ami33	33	1.27	1.29	1.22
Ami49	49	36.8	38.93	37.20
n30	30	0.247	0.234	0.227
n50	50	0.243	0.222	0.221

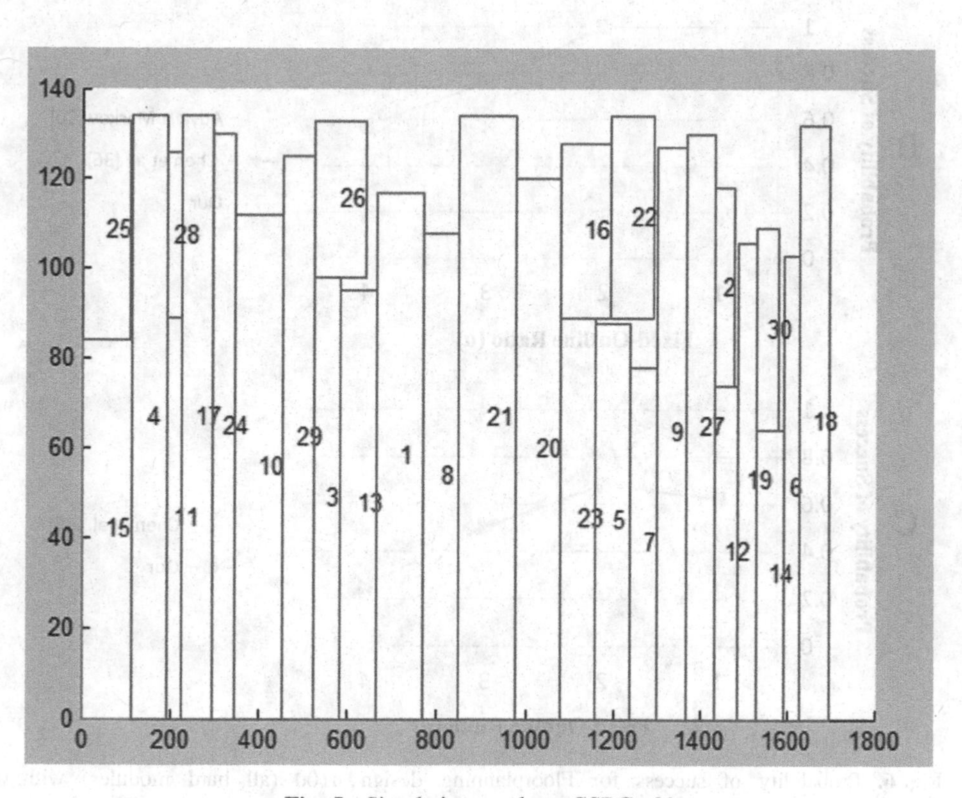

Fig. 5. Simulation result on GSRC n30

5.2 Floorplanning Under Fixed-Outline Constraints

In our formulation of the Floorplanning problem various assumptions are adopted from Adya and Markov (2003) [20]. For a given set of modules with total area A and given maximum percentage of dead space or wastage (Γ), a fixed outline is constructed with expected fixed outline ratio (aspect ratio) α, i.e. height/width. Equations (5), (6) mentioned above are used to calculate the values for height (H*) and width (W*), so

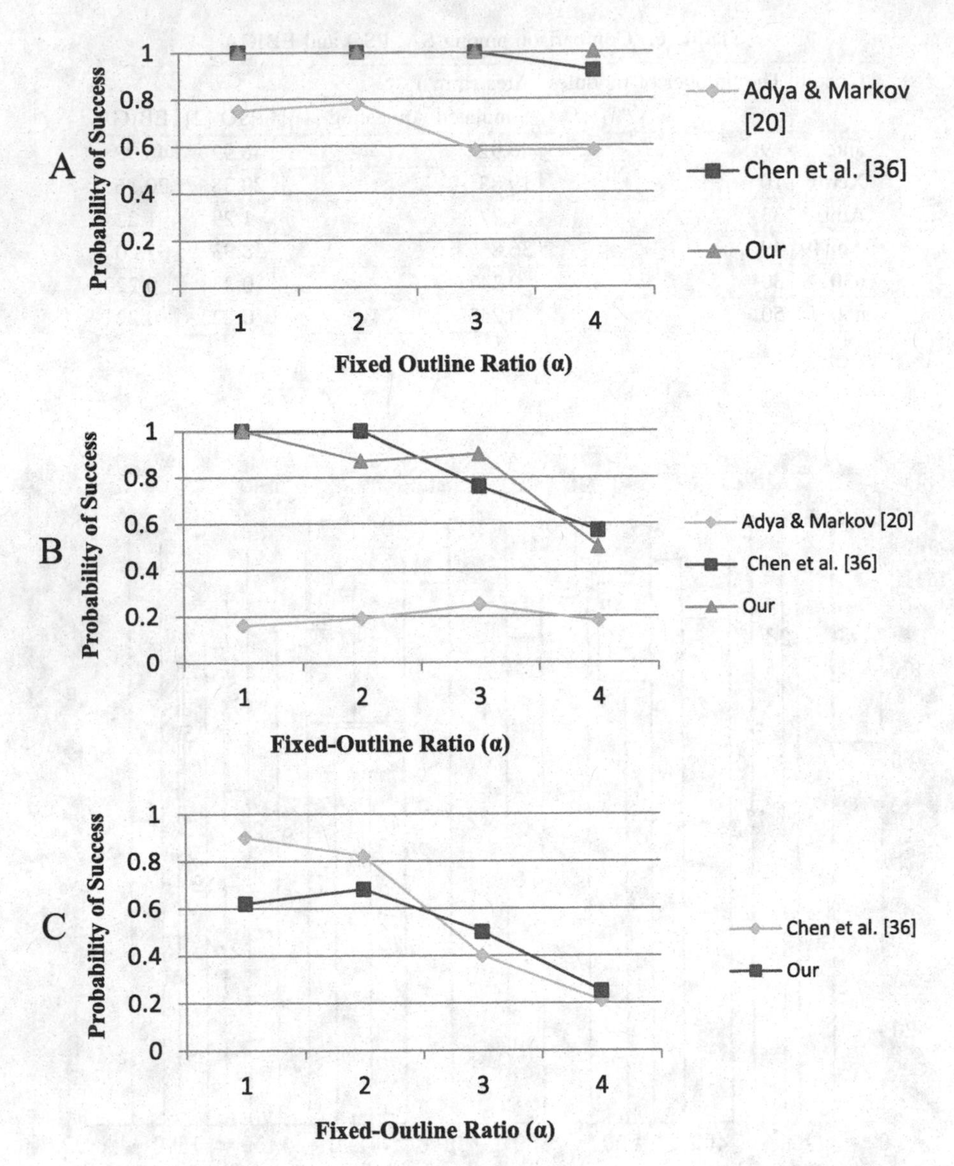

Fig. 6. Probability of success for Floorplanning design n100 (all hard modules) with fixed-outline constraints. The maximum dead-space (wastage) for the design is 15% i.e., $\Gamma = 15\%$ (Fig. A), 12%. i.e., $\Gamma = 12\%$ (Fig. B) and 10%. i.e., $\Gamma = 10\%$ (Fig. C). We plotted average of 50 runs for each fixed outline ratio (i.e. α = 1, 2, 3 and 4).

that the outline of floorplan can be determined. As soon as the solution satisfying a given fixed outline constraint is found proposed algorithm is terminated immediately, otherwise if no solution is satisfied and algorithm comes to an end, then it marks the evolution as a failure.

We have used the GSRC benchmark n100 (all hard modules) to conduct the experiment. The maximum value of dead-space (Γ) is set to 15%, 12% and 10%. Our results are compared with the results reported from Adya and Markov (2003) [20] and Chen et al. (2007) [35]. All our results are averaged for 50 runs for each outline ratio and runtimes are measured (in seconds) on a 1.7 GHz machine that runs Windows 7. Figure 6(a) show the plot for probability of success of satisfying the fixed outline constraints versus desired fixed-outline ratio with $\Gamma = 15\%$; and Fig. 6(b) show the plot for probability of success of satisfying the fixed outline constraints versus desired fixed-outline ratio with $\Gamma = 12\%$. Success rates of our algorithm are better than [20] by a big margin and are very competitive in comparison with [35], with reasonably increased run time. Figure 6(c) clearly shows that our algorithm is very competitive with [35] for $\Gamma = 10\%$. These experiments demonstrate that, the heuristic placement strategy proposed is more suitable for the fixed-outline problem than the slack-based approach proposed in Adya and Markov (2003) [20] and EBIGA provides much better results than Chen et al. (2007) [35] for some selected fixed outline ratio's. Comparison of average runtimes of our algorithm for Floorplanning design n100 with maximum dead-space (wastage) of ($\Gamma = 15\%$, $\Gamma = 12\%$ and $\Gamma = 10\%$) are provided in Fig. 7. Our approach can achieve floorplans satisfying different fixed-outlines with fairly good success rates and simulation results for dead-space ($\Gamma = 15\%$, $\Gamma = 12\%$ and $\Gamma = 10\%$) for various values of fixed outline ratio can be seen in Figs. 8, 9 and 10.

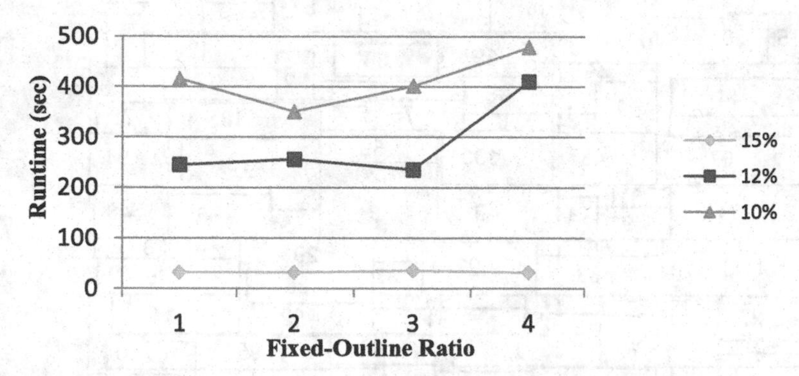

Fig. 7. Average runtimes of our algorithm for Floorplanning design **n100** with fixed-outline constraints and maximum dead-space (wastage) of $\Gamma = 15\%$, 12% and 10% are provided for comparison. We plotted average of 50 runs for each a fixed outline ratio (i.e. $\alpha = 1, 2, 3$ and 4).

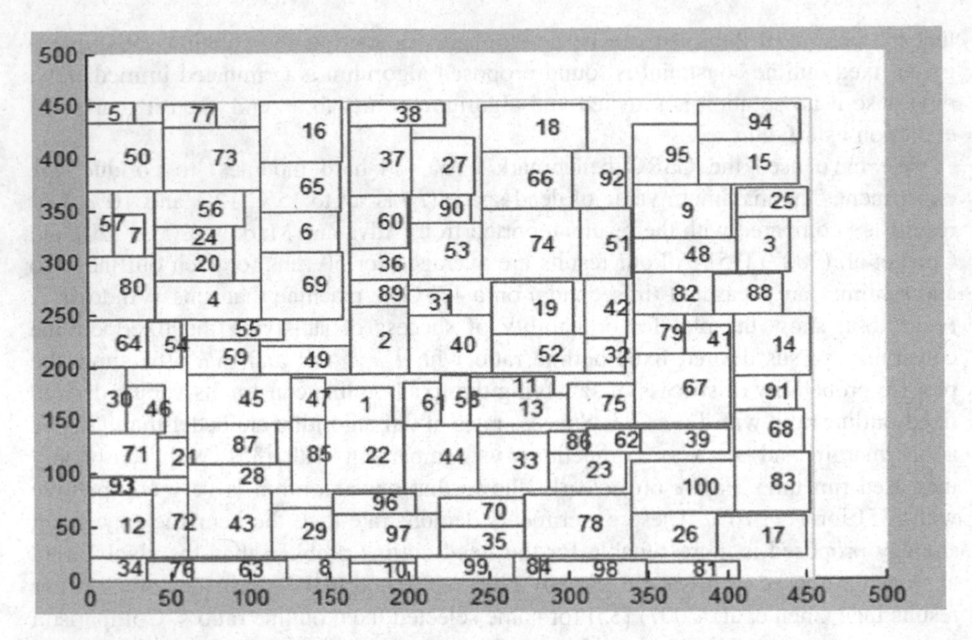

Fig. 8. Simulation Result on GSRC n100 for $\Gamma = 15\%$ and $\alpha = 1$

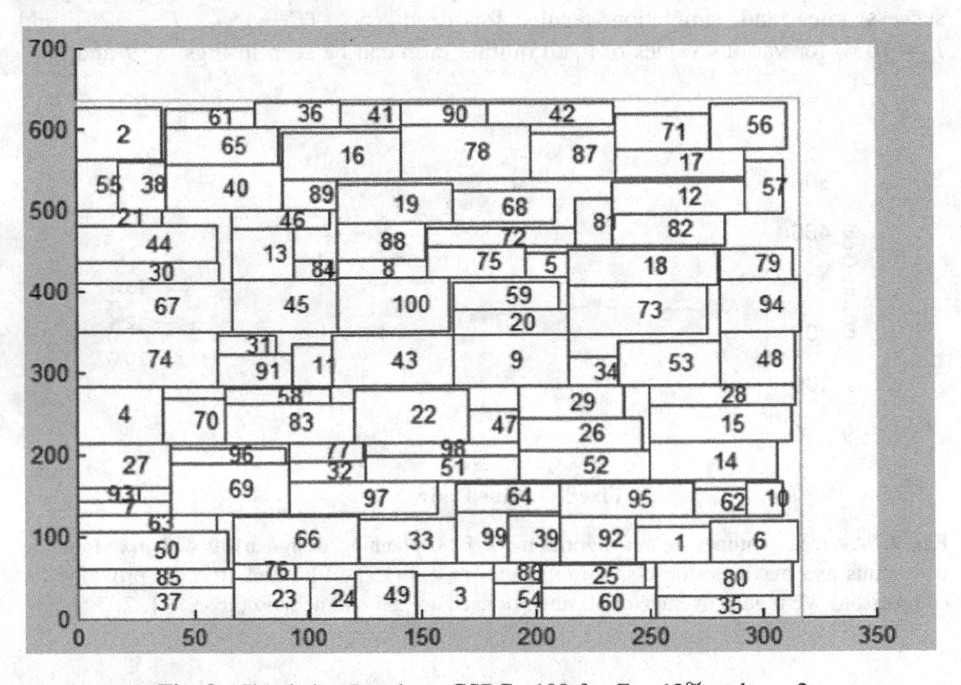

Fig. 9. Simulation Result on GSRC n100 for $\Gamma = 12\%$ and $\alpha = 2$

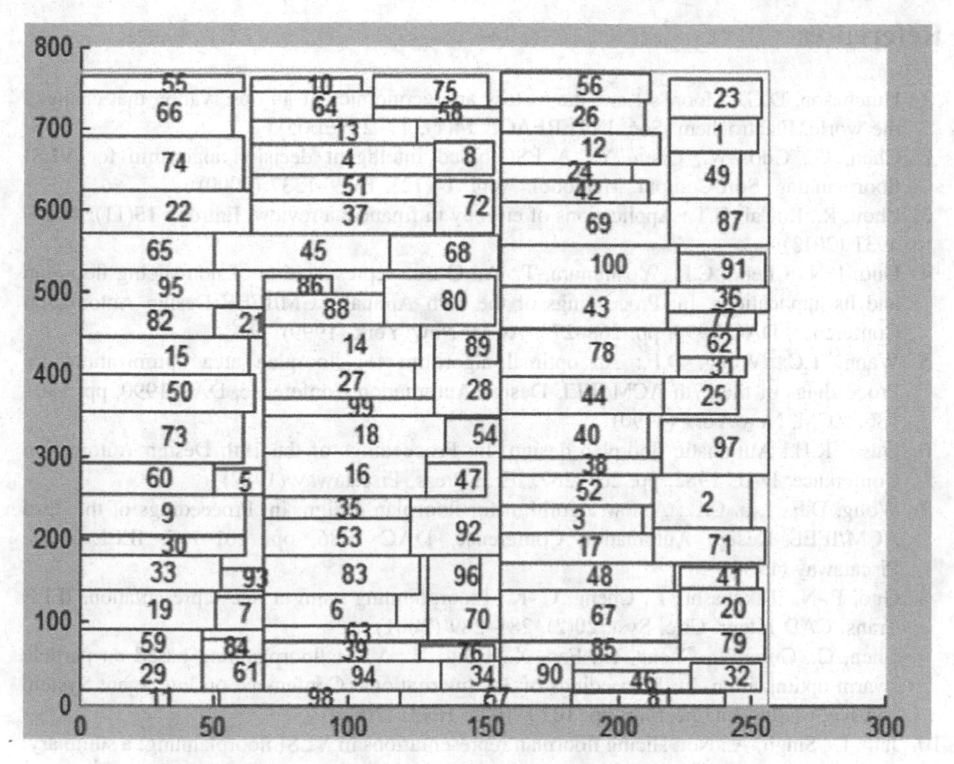

Fig. 10. Simulation Result on GSRC n100 for $\Gamma = 10\%$ and $\alpha = 3$.

6 Conclusion and Future Work

In this paper a robust evolutionary algorithm for modern VLSI floorplanning problem is presented. Highlight of our work lies in the proposed heuristic placement strategy and entropy based genetic algorithm having a crossover operation that helps to explore global search space in an efficient way. Different experiments show that our method achieves substantially improved success rate than most of other algorithms for both fixed outline and outline free floorplanning in an affordable time. Results for EBIGA show that various MCNC and GSRC benchmark circuits can be processed in a reasonable time. During the implementation of our work, it has been observed that the run time increase rapidly with the increase in number of modules. To overcome this problem, partitioning and multilevel techniques can be used for floorplanning. Another aspect that needs attention is to carry out optimization at other stages of physical design process of integrated circuits, such as Partitioning, Placement, Routing and Compaction.

Acknowledgement. Authors are thankful to the I.K. Gujral Punjab Technical University, Jalandhar for the support and motivation for research.

References

1. Hutcheson, D.G.: Moore's Law: the history and economics of an observation that changed the world. Electrochem. Soc. INTERFACE **14**(1), 17–21 (2005)
2. Chen, G., Guo, W., Chen, Y.: A PSO-based intelligent decision algorithm for VLSI floorplanning. Soft Comput. Methodol. Appl. **14**(12), 1329–1337 (2009)
3. Zhou, R., RuCai, G.T.: Applications of entropy in finance: a review. Entropy **15**(11), 4909–4931 (2013)
4. Guo, P.N., Cheng, C.K., Yoshimura, T.: An O-tree representation of non-slicing floorplan and its applications. In: Proceedings of the 36th Annual ACM/IEEE Design Automation Conference, DAC 1999, pp. 268–273. ACM, New York (1999)
5. Wang, T.C., Wong, D.F.: An optimal algorithm for floorplan area optimization. In: Proceedings of the 27th ACM/IEEE Design Automation Conference, DAC 1990, pp. 180–186. ACM, New York (1990)
6. Otten, R.H.: Automatic floorplan design. In: Proceedings of the 19th Design Automation Conference, DAC 1982, pp. 261–267. IEEE Press, Piscataway (1982)
7. Wong, D.F., Liu, C.L.: A new algorithm for floorplan design. In: Proceedings of the 23rd ACM/IEEE Design Automation Conference, DAC 1986, pp. 101–107. IEEE Press, Piscataway (1986)
8. Guo, P.-N., Takahashi, T., Cheng, C.-K.: Floorplanning using a tree representation. IEEE Trans. CAD Integr. Circ. Syst. **20**(2), 281–289 (2001)
9. Chen, G., Guo, W., Cheng, H., Fen, X., Fang, X.: VLSI floorplanning based on particle swarm optimization. In: Proceedings of 3rd International Conference on Intelligent System and Knowledge Engineering, pp. 1020–1025. IEEE (2008)
10. Jain, L., Singh, A.: Non slicing floorplan representations in VLSI floorplanning: a summary. Int. J. Comput. Appl. **71**(15), 12–19 (2013)
11. Hong, X.L., Huang, G., Cai, Y.C., Gu, J.C., Dong, S.Q., Cheng, C.K., Gu, J.: Corner block list: an effective and efficient topological representation of non-slicing floorplan. In: Proceedings of IEEE/ACM International Conference on Computer-Aided Design, pp. 8–12. ACM/IEEE (2000)
12. Murata, H., Fujiyoshi, K., Nakatake, S., Kajitani, Y.: VLSI module placement based on rectangle-packing by the sequence-pair. IEEE Trans. CAD **15**(12), 1518–1524 (1996)
13. Nakatake, S., Fujiyoshi, K., Murata, H., Kajitani, Y.: Module packing based on the BSG-structure and IC layout applications. IEEE Trans. CAD **17**(6), 519–530 (1998)
14. Guo, P.N., Cheng, C.K., Yoshimura, T.: An O-tree representation of non-slicing Floorplan and its applications. In: Proceedings of the 36th ACM/IEEE Conference on Design Automation, New Orleans, Louisiana, United States, pp. 268–273 (1999)
15. Lin, J.M., Chang, Y.W.: TCG: a transitive closure graph-based representation for non-slicing floorplans. In: Proceedings of the 38th Design Automation Conference, Las Vegas, USA, pp. 764–769 (2001)
16. Chang, Y.C., Chang, Y.W., Wu, G.M.: B*-tree: a new representation for non-slicing Floorplans. In: Proceedings of the 37th Conference on Design Automation, pp. 458–463. ACM, Los Angeles (2000)
17. Chen, T.-C., Chang, Y.-W.: Modern floorplanning based on B*-Tree and fast simulated annealing. IEEE Trans. Comput. Aid. Des. Integr. Circ. Syst. **25**(4), 637–650 (2006)
18. Gwee, B.H., Lim, M.H.: A GA with heuristic based decoder for IC floorplanning INTEGRATION. VLSI J. **28**(2), 157–172 (1999)
19. Kahng, A.B.: Classical floorplanning harmful. In: Proceedings of the 2000 International Symposium on Physical Design, ISPD 2000, pp. 207–213. ACM, New York (2000)

20. Adya, S.N., Markov, I.L.: Fixed—outline floorplanning: enabling hierarchical design. IEEE Trans. Very Large Scale Integr. (VLSI) Syst. **11**(6), 1120–1135 (2003)
21. Adya, S.N., Markov, I.L.: Fixed-outline floorplanning through better local search. In: Proceedings of the International Conference on Computer Design: VLSI in Computers & Processors, ICCD 2001, pp. 328–334. IEEE Computer Society, Austin (2001)
22. Kang, M., Dai, W.: Arbitrary rectilinear block packing based on sequence pair. In: Proceedings of IEEE/ACM International Conference on Computer Aided Design (ICCAD), pp. 259–266 (1998)
23. Tang, M., Yao, X.: A memetic algorithm for VLSI floorplanning. IEEE Trans. Syst. Man Cybernet. B Cybernet. **37**(1), 62–69 (2007)
24. Fernando, P., Katkoori, S.: An elitist non-dominated sorting based genetic algorithm for simultaneous area and wirelength minimization in VLSI floorplanning. In: International Conference on VLSI Design, pp. 337–342. IEEE (2008)
25. Rahim, H.A., Rahman, A.A.H. Ab., Andal jayalakshmi, G., Firuz, S.: A genetic algorithm approach to VLSI macro cell non-slicing floorplans using binary tree. In: Proceedings of the International Conference on Computer and Communication Engineering. IEEE (2008)
26. Chen, J., Zhu, W.: A hybrid genetic algorithm for VLSI floorplanning. In: International Conference on Intelligent Computing and Intelligent Systems (ICIS), pp. 128–132. IEEE (2010)
27. kiyota, K., Fuiiyoshi, K.: Simulated annealing search through general structure floorplans using sequence-pair. In: Symposium on Circuits and Systems, Geneva, Switzerland, pp. 77–80. IEEE (2000)
28. Fang, J.-P., Chang, Y.-L., Chen, C.-C., Liang, W.-Y., Hsieh, T.-J., Satria, M.T., Han, C.-C.: A parallel simulated annealing approach for floorplanning in VLSI. In: Hua, A., Chang, S.-L. (eds.) ICA3PP 2009. LNCS, vol. 5574, pp. 291–302. Springer, Heidelberg (2009). https://doi.org/10.1007/978-3-642-03095-6_29
29. Anand, S., Saravanasankar, S., Subbaraj, P.: Customized simulated annealing based decision algorithms for combinatorial optimization in VLSI floorplanning problem. Comput. Optim. Appl. **52**(3), 667–689 (2011)
30. Chen, J., Zhu, W., Ali, M.M.: Hybrid simulated annealing algorithm for nonslicing VLSI floorplanning. IEEE Trans. Syst. Man Cybernet. Part C Appl. Rev. **41**(4), 544–553 (2011)
31. Chen, Z., Chen, J., Guo, W., Chen, G.: A co-evolutionary multi-objective PSO algorithm for VLSI floorplanning. In: 8th International Conference on Natural Computation (ICNC), pp. 712–728. IEEE (2012)
32. Wang, X.G., Yao, L.S., Gan, J.R.: VLSI floorplanning method based on genetic algorithms. Chin. J. Semiconductors **23**, 330–335 (2002)
33. Pang, Y., Cheng, C.K., Yoshimura, T.: An enhanced perturbing algorithm for floorplan design using the O-tree representation. In: Proceedings ISPD, pp. 168–173 (2000)
34. Tang, X., Wong, D.F.: FAST-SP: a fast algorithm for block placement based on sequence pair. In: Proceedings on ASPDAC (2001)
35. Chen, D.-S., Lin, C.-T., Wang, Y.-W., Cheng, C.-H.: Fixed-outline floorplanning using robust evolutionary search. Eng. Appl. Artif. Intell. **20**, 821–830 (2007)

Enhancing Levenshtein's Edit Distance Algorithm for Evaluating Document Similarity

Shama Rani and Jaiteg Singh(✉)

Chitkara University, Chandigarh, India
shamajhansla@gmail.com, Jaiteg.singh@chitkara.edu.in

Abstract. The content directly taken from pre-published sources is called plagiarized text. Plagiarism is considered to be a major challenge in contemporary research manuscripts. It is very easy to use internet as a source of information. There is a need to find suitable technique, so as to find similarity between two documents. Though there are several methods for text comparison, yet this paper is primary focused on Levenshtein's edit distance. It is a string metric, which is an effective technique for comparing text documents and for calculating efforts required to transforms one document to another. An effort has been made to improve the performance of Levenshtein's edit distance algorithm by eliminating stop words while calculating transformation effort.

Keywords: Comparison of documents · Levenshtein's edit distance
Modified Levenshtein's edit distance · Similarity between documents
Cost and time evaluation

1 Introduction

Plagiarism is defined as the use of another's thoughts, literature, and information, when done without proper citation of the original source. Plagiarism for the text documents occurs in different ways. Plagiarized text may be copied from one-to-one passages may be modified to a larger or reduced extent or text may be interpreted [15]. Data Comparison relates to the methods of calculating differences and similarities so to replace the strings and data objects. The objects that are compared usually program code, algorithms, computer files, text versions, or complex data structures [16].

To detect plagiarism in software presents some problems due to the nature of programming. The reasons for similarity between the programs can be categorized in different categories, one of which is plagiarism. Similarities including metric, textual, features in depth and some recommendations are made for measures of syntax and semantics, program execution, input-output, shared information, program dependency graph similarity [3, 4].

There are two most probable features used to compare documents are: importing one single file for online plagiarism check, or matching two file for a comparative check. It follows the following steps:

Step1. The text is exported from the file with ignorance of images and other diagrams.

Step 2. The text is divided into n –grams or sets of words.

R. Sharma et al. (Eds.): ICAN 2017, CCIS 805, pp. 72–80, 2018.
https://doi.org/10.1007/978-981-13-0755-3_6

Step 3. Each group is searched by the software.

Step 4. The search engine results are stored and loaded on pages.

Step 5. The page is parsed when the website has been loaded to extract the text from the HTML code.

Step 6. The string is explored inside the mined text.

Step 7. If a matching sentence has been found, the input text is added to the source list and the next Sentence is starting to be analyzed.

Step 8. If the sentence has not been found, another website is loaded until results were analyzed. The two algorithms used for the detection of Text is Levenshtein's Edit distance method [5, 13].

2 Levenshtein's Edit Distance

The Levenshtein's distance between two documents is defined as the minimum number of edit operations required to transform one text document into the other. The following edit operations are used by Levenshtein's edit distance algorithm to modify one document to another:

- Insertion
- Deletion
- Substitution

Example: Levenshtein's edit Distance between different strings:
right → fight (substitution of 'f' for 'r')
book → books (insert operation is performed at the end 's')

The Levenshtein's edit Distance between given strings depends on three basic operations to replace one string to another.

The results of Levenshtein's distance is based on the perception that if a matrix holds the edit distance between all prefixes of the first string and all the prefixes of the second, Thus find the distance between the two full strings as the last value calculated [6, 10].

The algorithm:

Step 1: Initialization

I. Set a is the length of document 1 say d1, set b is length of document 2 say d2.
II. Create a matrix that consists 0 – b rows and 0 – a columns.
III. Initialize the first row from 0 to a.
IV. Initialize the first column from 0 to b.

Step2: Processing

I. Observe the value of d2 (i from 1 to a).
II. Observe the value of d1 (j from 1 to b).
III. If the value at d2[i] is equals to value at d1[j], the cost becomes 0.
IV. If the value at d2[i] does not equal d1[j], the cost becomes 1.
V. Set block of matrix M[d1, d2] of the matrix equal to the minimum of:
i. the block immediately above add 1: M [d2–1, d1]

ii. the block immediately to the left add 1: M [d2, d1–1] +1.
iii. The block is diagonally above and to the left adds the cost: M [d2–1, d1–1] + cost.

Step 2 is repeated till the distance M[a, b] value is found.
Step 3: Result [1, 5].

2.1 Computing Techniques

- Dis(i,j) = score of best alignment from d11..d1i to d21…..d2j
- Dis (i–1, j–1) + d (d1i, d2j) //copy Dis (i–1, j) + 1 //insert
- Dis (i, j–1) + 1 //delete

Cost depend upon following factors:

- Dis (0, 0) = 0 cost // if both strings are same
- Dis (i, 0) = dis (i–1, 0) +1 = 0 // if source string is empty
- Dis (0, j) = dis (0, j–1) +1 = 0 // if target string is empty [6, 7].

3 Modified Levenshtein's Edit Distance Algorithm

Levenshtein's edit distance algorithm can be modified by removing the stop words. The words such as also, is, am, are, they, them, their, was, were etc. are ignored by search engine during processing are called stop words [4, 8].

Most search engines are programmed to eliminate such words while indexing or retrieving as the outcome of search query. Stop words are considered inappropriate for searching purposes because they occur commonly in the language for which the indexing engine has been tuned [4]. These words are dropped in order to save both time and space at the time of searching in the text documents. The words which are often used are is, am, are, they, them, also, the, of, and, to, in which are insignificant in IR and text mining. Stop words are removed to reduce due to following reasons:

- Each documents approximately consists 20–25% stop words.
- Efficiency of document is improved by removing the stop words.
- Stop words are not useful for searching or text mining
- To reduce indexing [9].

4 Performance Analysis

Levenshtein's edit Distance is not considered as an absolute value. If the first string is 'Race' and the second string is 'spaces', it's very unlikely that one of them is misspelled. However, if the first string is 'I have a pet' and the second string is 'I have a cat', the second string is probably misspelled. But in both cases, the Levenshtein's Distance is 2. The first value means that 2/3 of the characters are different, the second value tells us that the difference is little. Another method to calculate Levenshtein's edit distance algorithm is matrix method. The Levenshtein's Edit Distance algorithm calculates the minimum edit operations that are needed to modify one document to obtain second document. A matrix is initialized measuring in the (m, n)-cell the Levenshtein's distance between the m-character prefix of one with the n-prefix of the other word [12, 13].

The following example will determine the use of Matrix method. Let the first string is PEON and second string is SPEND the minimum path is selected by comparing at each stage. The calculation process of the Levenshtein's distance between two strings of different length is based on the number of operations to transform first string to another and the edit distance between the substrings X1m = x1x2…xn and Ymn = y1 y2…yn is calculated as follows:

- Dis(m,n) = Dis(X1m, Y1n)
- $\mathrm{Dis(m,n) = min\{Dis(m-1,n)+1, Dis(m,n-1)+1, Dis(m-1,n-1)+Cost\}}$

$$\text{With } 0 \quad \text{if } Xm-1 = Yn-1$$
$$1 \quad \text{else}$$

and the initializations are: $D(m,\varnothing) = m$ and $D(\varnothing,n) = n$, where $\varnothing$ represents the empty string [11].

In this way, we calculate the Levenshtein's distance between two strings is shown in lower right most block in Fig. 1 [13]. In above example there are different ways to replace "PEON" with "SPEND", but the minimum cost path is taken by this method is shown with arrows. The experimental details of Levenshtein's Edit distance in terms of space and time is follows. The inputs given in two Documents and number of words after removing stop words are calculated in all the documents by using Levenshtein's Edit distance formula is shown in Fig. 2. The time taken to calculate Levenshtein's distance with Stop words is shown in Fig. 3. The calculated time and cost represented in separately in view of easy understanding of graphs. The time taken to compare documents is calculated in milliseconds and later be converted into asymptomatic time by applying on some other algorithms.

		P	E	O	N
	0	1	2	3	4
S	1	1 ↓	2	3	4
P	2	1	2	3	4
E	3	2	1 →	2 →	3
N	4	3	2	2	2 →
D	5	4	3	3	3 ↓

Fig. 1. Edit distance between two strings

- D(m,n) = score of best alignment from s1..si to tm……tn.

Here 51 Text length of Document A and B means the document size. The document size means it contains the defined number of words. The experiment is done by taking different document size from 50–1000.

Text Length of Document A	Text Length of Document B	Document A after removing stop words	Document B after removing stop words
51	62	27	38
103	90	59	53
203	192	124	119
395	410	242	233
798	750	470	474

Fig. 2. Text length of Document A and B with and without using stop words

Text Length of Document A	Text Length of Document B	Time taken to calculate Levenshtein's distance with Stop words (in milliseconds)
51	62	14
103	90	16
203	192	23
395	410	62
798	750	218

Fig. 3. Time taken to calculate Levenshtein's distance after removing stop words

The time taken to calculate Levenshtein's distance with Stop words is represented in Fig. 3. The Time taken to calculate Levenshtein's distance after removing Stop words is shown in Fig. 4.

The length of Document A and the length of same document after removing the stop words is shown in Fig. 5. Where case A1 represents the complete text length of Document A and case A2 represents the length of Document A after removing the stop words.

Similarly, in Fig. 6. case, B1 represents the complete text length of Document B and case B2 represents the length of Document B after removing the stop words. Of the Levenshtein's distance between the words "PEON" and "SPEND", the distance is three as shown in Fig. 1.

The matrix completes the blocks from the top most corner of left to the lower right corner. Each move vertically or horizontally corresponds to insertion or a deletion and substitution respectively. Each operation is initially set to costs.

1. The diagonal move costs one, if the two characters in the row and do not match and one if they do. The cost is locally minimizes by each block.

Text Length of Document A	Text Length of Document B	Time taken to calculate Levenshtein's distance after removing Stop words (in milliseconds)
51	62	7
103	90	5
203	192	12
395	410	30
798	750	119

Fig. 4. Time taken to calculate Levenshtein's distance after removing stop words

- case TW1 represent the time with stop words and TWO represents the time taken to calculate Levenshtein's distance without stop words.

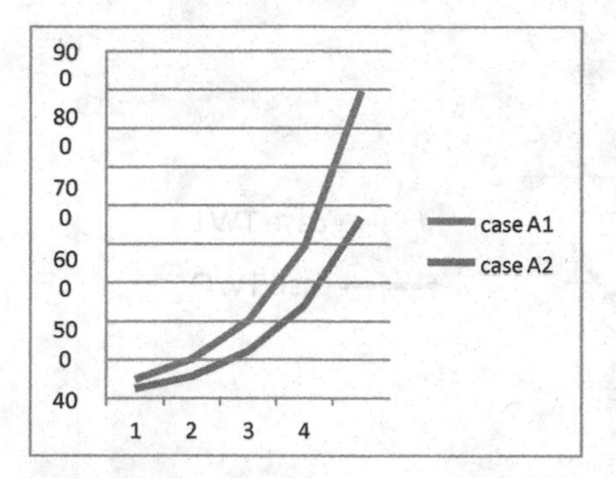

Fig. 5. Text length of document A with and without using stop words

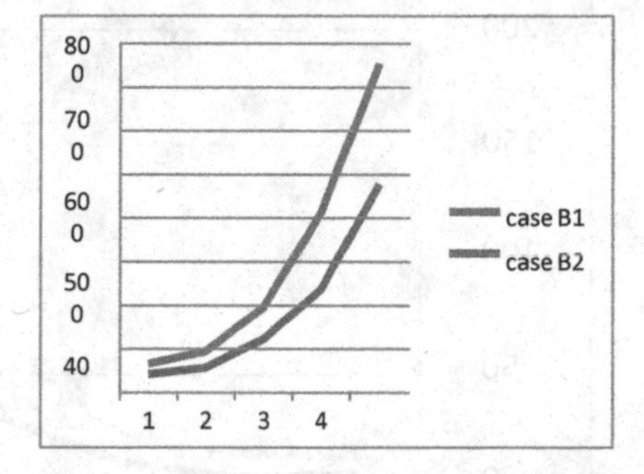

Fig. 6. Text length of document B with and without using stop words

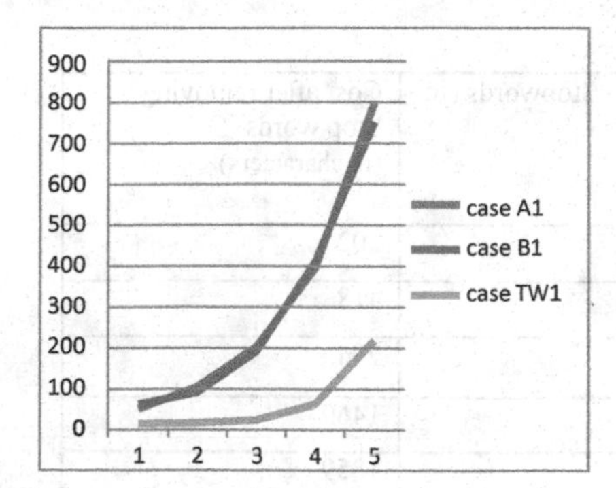

Fig. 7. Time taken to calculate Levenshtein's distance with stop words

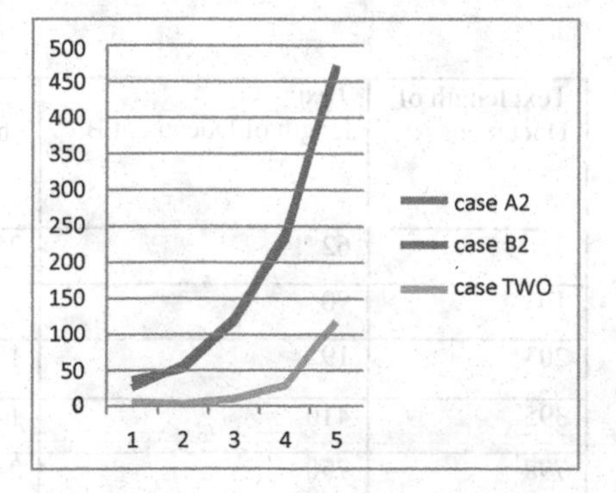

Fig. 8. Time taken to calculate Levenshtein's distance without stop words

The time taken to calculate Levenshtein's edit distance with Stop words is shown in Fig. 7. where Case A1 represents the complete text length of Document A and case B1 represents the complete text length of Document B. The case TW1 is the time taken by algorithm to compare the lengths of Document A and Document B.

The Fig. 8. represents the comparison of Documents after removing the Stop words. Case A2 signifies the length of Document A after removing the stop words, case B2 represents the length of Document B after removing the stop words and The final comparison of the times taken by Levenshtein's in different documents with and without stop words. In Fig. 9. case TW1 is time taken to calculate Levenshtein's distance with stop words and case TWO is time taken to calculate edit distance after removing the stop words in the text Documents.

The cost calculated by Levenshtein's Algorithm of two dissimilar documents is shown in Fig. 10.

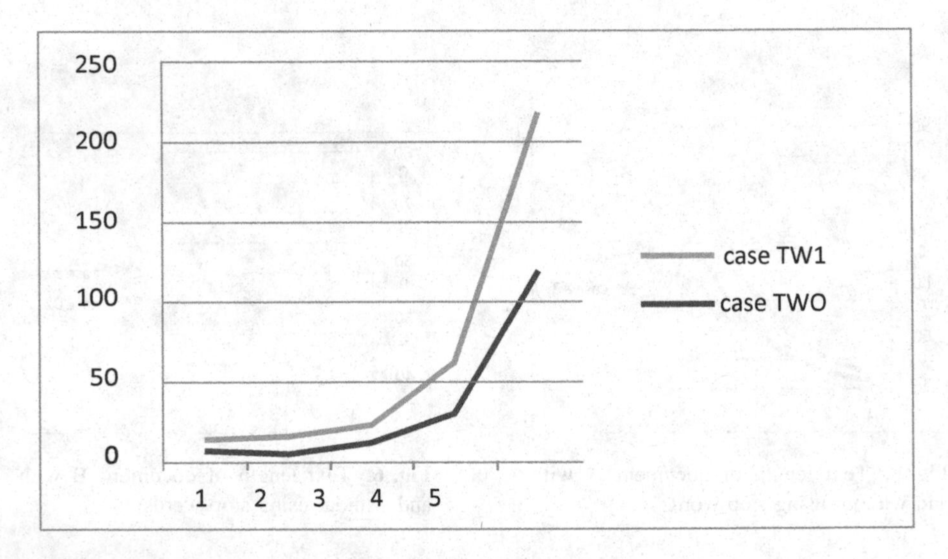

Fig. 9. Time taken to calculate before and after removing stopwords

Text length of Document A	Text length of Document B	Cost with Stopwords (in characters)	Cost after removing Stop words (in characters)
51	62	257	202
103	90	534	418
203	192	1011	794
395	410	1907	1469
798	750	2483	1859

Fig. 10. Cost to replace one document to another

5 Conclusion

The documents with different text length 50, 100, 200, 400, 800 is taken to calculate the Levenshtein edit distance and the time need to compare both documents by using Levenshtein edit distance algorithm. This is observed that each document consists 20–30% stop words, which are not useful for any calculation. Therefore, it is observed that if 20% stop words are removed from any text document, 50% time can be reduced to calculate the Levenshtein's edit distance. The Comparison of time with stop words and after removing stop words from the documents of different text length is shown in Fig. 9.

References

1. Gueddah, H., et al.: Introduction of the weight edition errors in the Levenshtein distance. IJARAI Int. J. Adv. Res. Artif. Intell. **1**(5) (2012)
2. Dang, Q.T.: Determining restricted Damerau-Levenshtein edit-distance of two languages by extended automata. In: International Conference on Computing and Communication
3. Burkhardt, S., Kärkkäinen, J.: One-Gapped q-Gram filters for Levenshtein distance. In: Apostolico, A., Takeda, M. (eds.) Combinatorial Pattern Matching. CPM 2002. Lecture Notes in Computer Science, vol. 2373, pp. 225–234. Springer, Heidelberg (2002). https://doi.org/10.1007/3-540-45452-7_19
4. Hubert, C.: A contextual normalized edit distance. In: 2008 IEEE 24th International Conference on Source Data Engineering Workshop, ICDEW 2008 (2008)
5. Aouragh, S.I.: Adaptating Levenshtein distance to contextual spelling correction. Int. J. Comput. Sci. Appl. **12**(1), 127–133 (2015)
6. Danny, H.: A unified algorithm for accelerating edit distance computation via text compression. In: Symposium on Theoretical Aspects of Computer Science year (city)
7. Ndiaye, M., Faltin, A.V.: Correcteur Orthographique Adapté à Apprentissage du Français. Revue Bulag no. 29, pp. 117–134 (2004)
8. Mitton, R.: Ordering the suggestions of a spellchecker without using context. Nat. Lang. Eng. **15**(2), 173–192 (2009)
9. Damerau, F.J.: A technique for computer detection and correction of spelling errors. Commun. Assoc. Comput. Mach. **7**, 171–176 (1964)
10. Kobzdej, P.: Parallel application of Levenshtein's distance to establish similarity between strings. Front. Artif. Intell. Appl. **12**(4) (2003)
11. Yujian, L., Bo, L.: A normalized Levenshtein's distance metric. IEEE Trans. Pattern Anal. Mach. Intell. **29**(6), 1091–1095 (2007)
12. Haldar, R., Mukhopadhyay, D.: Levenshtein distance technique in dictionary lookup methods: an improved approach. Web Intell. Distrib. Comput. Res. Lab
13. Andoni, A.: Approximating edit distance in near linear time. In: Proceedings of the 41st Annual ACM Symposium on Theory of Computing (STOC 2009), Bethesda, MD, USA, pp. 199–204 (2009)
14. Oberreuter, G., Velasquez, J.D.: Expert system with applications: text mining applied to plagiarism detection. Web Intelligence Consortium Chile Research Centre, Department of Industrial Engineering, Chile (2013)
15. Kharat, R., et al.: Semantically detecting plagiarism for research papers. Int. J. Eng. Res. Appl. (IJERA) **3**(3), 077–080 (2013). ISSN: 2248-9622, https://www.ijera.com

16. Caroline, L., et al.: Plagiarism is Easy, But also Easy to Detect. MPublishing, University of Michigan Library, Ann Arbor, vol. 1 (2006)
17. Journal of technology management for growing economies (JTMGE) (2015). http://tmgejournal.com/abstract.php?id=513. Accessed 14 Jan 2017

Analytics

Connected Cluster Based Node Reduction and Traversing for the Shortest Path on Digital Map

Dhaval Kadia(✉)

The Maharaja Sayajirao University of Baroda, Vadodara 390001, Gujarat, India
dhavalkadiamsu@gmail.com

Abstract. This paper elaborates the algorithm and methods applied on a digital map and explains how the digital map is reduced into its nodal form for obtaining the shortest path along with navigation, directions and the individual distances. The algorithm uses different types of data structures appropriate for different approaches and interacts with the digital map or its nodal form. Auxiliary data structures along with the respective methods are added in the algorithm to decrease the time and space complexity. The time-memory tradeoff is observed during the implementation through the two memory allocation schemes. In the beginning, static memory allocation is used which is followed by the dynamic memory allocation. Both the schemes have their corresponding consequences. Precision in navigation is varied on the basis of peculiar requirements so that the overall complexity of the algorithm can be reduced. The algorithm first processes the digital map and identifies roads and junctions which will be the input for further processing. Cluster based approaches are applied for simplification and efficiency purposes. The paper compares the various algorithms for obtaining the shortest path. Comparison of the diverse experiment results show the improvement in the execution time and memory consumption.

Keywords: Shortest path · Algorithm · Analysis · Data structure
Cluster · Node representation · Node reduction · Optimization
Digital map · Navigation

1 Introduction

The shortest path algorithm is conceptualized on an appropriate data structure. Digital map, an image having numeric values, represents roads. The road is considered to be present where the connected pixels have 1 s. Digital map is a network that is taken as an input for finding junctions within it. Program memory is allocated on static or dynamic basis and as per starting and destination points, the algorithm is then executed, data structures are manipulated and decisions are taken. Initially, the paper explains the basic algorithm and its interaction with the data structure. Then, it describes how optimization is applied on the data structure as well as on the algorithm. Subsequently, it demonstrates how auxiliary data structures are introduced. These three phenomena are elaborated in their respective sections. The entire algorithm, along with its further

R. Sharma et al. (Eds.): ICAN 2017, CCIS 805, pp. 83–96, 2018.
https://doi.org/10.1007/978-981-13-0755-3_7

optimization, is implemented in Language C. Sufficient care is taken on memory deallocation by applying the methodologies present in [1, 4].

2 Initialization

Throughout the paper, the two-dimensional array *(MAP)* refers to the digital map. Different colors in figures are associated with their values of pixels and behaviors. L_s and L_d symbolizes starting location and destination location respectively. They are represented in terms of the Cartesian coordinate system i.e. (x_{start}, y_{start}) and $(x_{destination}, y_{destination})$. Digital map represents the interconnected paths from which the algorithm gets reference of available paths and locations from where there may be more than one way to go further i.e. a junction. Initially, a digital map is a binary-valued *MAP*.

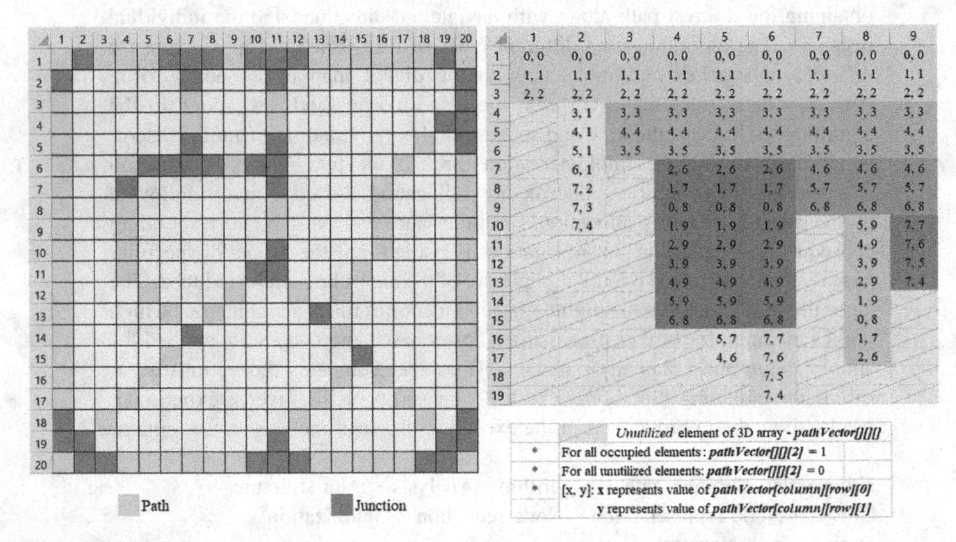

	1	2	3	4	5	6	7	8	9
1	0, 0	0, 0	0, 0	0, 0	0, 0	0, 0	0, 0	0, 0	0, 0
2	1, 1	1, 1	1, 1	1, 1	1, 1	1, 1	1, 1	1, 1	1, 1
3	2, 2	2, 2	2, 2	2, 2	2, 2	2, 2	2, 2	2, 2	2, 2
4		3, 1	3, 3	3, 3	3, 3	3, 3	3, 3	3, 3	3, 3
5		4, 1	4, 4	4, 4	4, 4	4, 4	4, 4	4, 4	4, 4
6		5, 1	3, 5	3, 5	3, 5	3, 5	3, 5	3, 5	3, 5
7		6, 1		2, 6	2, 6	2, 6	4, 6	4, 6	4, 6
8		7, 2		1, 7	1, 7	1, 7	5, 7	5, 7	5, 7
9		7, 3		0, 8	0, 8	0, 8	6, 8	6, 8	6, 8
10		7, 4		1, 9	1, 9	1, 9		5, 9	7, 7
11				2, 9	2, 9	2, 9		4, 9	7, 6
12				3, 9	3, 9	3, 9		3, 9	7, 5
13				4, 9	4, 9	4, 9		2, 9	7, 4
14				5, 9	5, 9	5, 9		1, 9	
15				6, 8	6, 8	6, 8		0, 8	
16					5, 7	7, 7		1, 7	
17					4, 6	7, 6		2, 6	
18						7, 5			
19						7, 4			

Fig. 1. Junction identification in digital map; 2D representation of 3D array *pathVector[ ][ ][ ]* having values assigned for $(x_{start}, y_{start}) = (1, 1)$ and $(x_{destination}, y_{destination}) = (8, 5)$. (Color figure online) (Figure is generated by retrieving values from data structures residing in the memory while execution).

Figure 1 shows pixels representing paths and junctions having values 1 and 2 respectively. Procedure for junction identification is now explained. $X - axis$ is vertical and $Y - axis$ is horizontal. That is, row and column numbers represent values on $X - axis$ and $Y - axis$ respectively. We as humans can identify junctions by looking directly to the digital map but for the algorithm to proceed further, it is necessary to determine locations where there are junctions. The function which is designed to find junctions, is now explained. As explained earlier, all coordinates having path will have value 1. The junction is a coordinate where the present path splits into two or more paths. So, current coordinate will be a junction if it is connected with more than two coordinates. It will have more than two adjacent pixels having value 1. The function *identifyJunction()*

performs the required task as it periodically checks all the coordinates and its adjacent coordinates. As a result, if it has more than two adjacent coordinates having values 1 or 2, then it will be assigned value 2. Thus, the algorithm will make its decision whether to split or not during the execution.

3 Static Memory Based Algorithm (SMA)

Static Memory based Algorithm *(SMA)* runs on the allocated memory prior to runtime. All variables are of type *Integer*. It finds all the possible paths and saves all the coordinates sequentially in 3D data structure while execution. These procedures are described in the upcoming subsections in detail. Working of *SMA* is elaborated using the basic example. Figure 2 represents an example to demonstrate the outcome of function *identifyJunction(),* through which *SMA* will execute. Input parameters to the algorithm are L_s and L_d. Algorithm performs functions shown in Fig. 3 on each path and junctions and assigns values demonstrated in Fig. 1 into a data structure. At the end, it analyzes them and finalize the shortest path shown in Fig. 2.

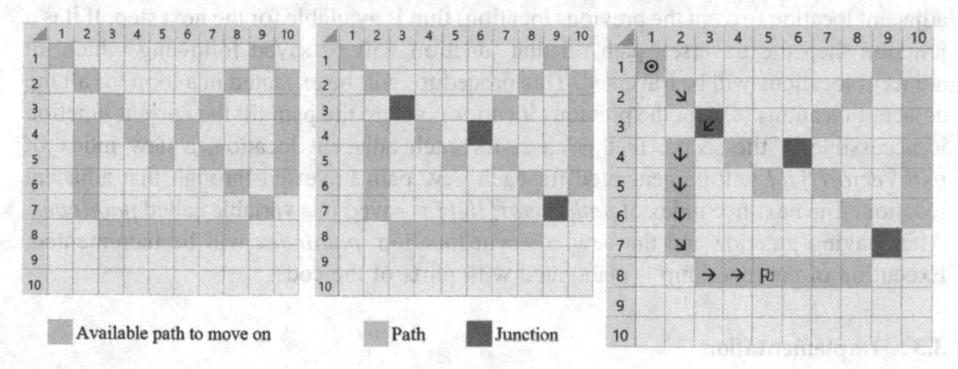

Fig. 2. Digital map to elaborate the *SMA* approach; Output of the shortest path between L_s = (1, 1) to D_s = (8, 5); Generated data structure: *pathVector[2][][]* in Fig. 1.

3.1 Data Structure

Data structure *pathVector[][][]* is a 3D array that stores all the possible paths from L_s to L_d. Its significance is discussed now. *pathVector[index][state][]*: *index* and *state* as column and row respectively in Fig. 1, *pathVector[][][0]* = x coordinate of any location, *pathVector[][][1]* = y coordinate of any location, *pathVector[][][2]* = 1 if (x, y) is a path else 0. Initially, L_s is inserted into *pathVector[1][1][]* i.e. x_{start}, y_{start} into *pathVector[1][1][0]* and *pathVector[1][1][1]* respectively.[1] Data structure *found[]* is an integer array to store all the indexes of *pathVector[index][][]* in which the last *state*

[1] Index 0 is shifted into index 1 only for first two dimensions so that representation can be user friendly i.e. *pathVector[0][][]* and *pathVector[][0][]* are shown as *pathVector[1][][]* and *pathVector[][1][]* only in figures. Index starts from 0 in actual practical implementation in Language C.

has L_d. That is $pathVector[index][state_{last}][0] = x_{destination}$ and $pathVector[index][state_{last}][1] = y_{destination}$. Till $state = state_{last}$, $pathVector[index][state_{last}][2]$ is 1. For next state = $state_{last}$+ 1, $pathVector[index][state_{last}+ 1][2]$ is 0, indicating that the ongoing path is now terminated. The *foundIndex* is a variable pointing to the current index of array *found[]* where the upcoming index value will be stored and the terminating location is L_d. This data structure is modified further for faster execution.

3.2 Algorithm Description

The algorithm starts its execution from L_s and terminates at L_d. Since L_s is inserted into *pathVector[][][]* as stated earlier, it is the location from where the algorithm explores all the possible paths through its adjacent locations towards L_d. This procedure is repeated continuously. Recursion is responsible for finding the shortest path automatically but for this, current paths must be saved so that they can be compared with upcoming paths and hence, the shortest distance can be computed. Algorithm saves coordinates of the path into *pathVector[][][]* on which it is currently moving. Now, for each next adjacent location, there are two possibilities i.e. whether it will be a path or a junction. If it is a path then the algorithm simply keeps moving on it, which indicates that there is only one adjacent location (except the previous location) that is available for the next step. If it is a junction then the traversed path till that junction will be saved following which all adjacent locations will be traversed. This procedure will be executed in a loop for all the adjacent locations (except the previous location), where the path till the current junction is accessible in the scope. In this case, for each adjacent location, a new index of *pathVector[][][]* will be generated for each new path traversed through that adjacent location. The next free index of *pathVector[][][]* is saved in a variable called *pathIndex*. After having junction and the new adjacent location, *pathIndex* will be incremented. Execution of the algorithm is elaborated with parts of the code.

3.3 Implementation

Functions *trace()*, *split()*, *getEndpoint()* and *replicateIndex()* are explained below using the definition and the body of each. The code is having pointers of Language C. Pointers are used so that they can be parameterized and passed into the function representing the address of the particular 2D array.

```
void trace(int x, int y, int **p, int vect)
{
    if (x, y) is Destination location
        found[foundIndex] = vect
        foundIndex++
    else
        if (x, y) is Junction
            split(x, y, p, vect)
        else if (x, y) is Path
            for each adjacent location (m, n) of (x, y)
                if getEndpoint(m, n, p) is false
                    if MAP[m][n] is Path
                        insert (m, n) at p[branchEnd][]
                        trace(m, n, p, vect)
                    else if MAP[m][n] is Junction
                        trace(m, n, p, vect)
}
```

```
void split(int x, int y, int **p, int vect)
{
    insert (x, Y) at p[branchEnd][]
    for each adjacent location (m, n) of (x, y) do
        if(MAP[m][n]> 0 && !getEndpoint(m, n, p))
            newIndex = replicateIndex(vect)
            insert (m, n) at pathVector[newIndex][branchEnd][]
            trace(m, n, pathVector[newIndex], newIndex)
}
```

Fig. 3. Definitions and bodies of functions *trace()* and *split()*. (End of *if* and *for* is not shown in figures because their scopes are indented by their relative positions).

Function Trace(). Algorithm executes the function shown in Fig. 3 each time it moves on to the next adjacent location. If the current location is L_d then, the currently used index of *pathVector[index][][]* will be saved as *found[foundIndex] = index* where *foundIndex* will be incremented. Thus, the list of paths having L_d as a terminating location can be acquired from *found[]* for further processing. If the current location is a junction then it will call the function *split()*. Else, if it is a path then it will find the next adjacent location. If the adjacent location is a path then it will insert that adjacent location into the end location of the current branch i.e. at the location pointed by *branchEnd*, which gives the array index that is not filled by the location value earlier and comes next to that array index having a location value. Here, ***p* is a pointer of a 2D array to the *pathVector[][][]* which comes from parameters of the function. Function *trace()* is called by inserting that adjacent location as a parameter, ***p* being same as that of the parameters. Parameter *vect* points the *pathIndex* of ***p*. If an adjacent location is a junction then the function *trace()* will be called without any insertion and control will be transferred to the function *split()*.

Function Split(). The function shown in Fig. 3 is applied when the current location is a junction. Since it is a junction, there will be more than one possibility where the algorithm will spread itself. Those new possible paths will have paths from start till the current location in common. After the initialization of these common locations, their independent path locations will be filled up in arrays pointed by their respective indexes. Here, the function *replicateIndex(int)* returns an *Integer* value of the array index that is available for replicating common location coordinates. Current *pathIndex* is pointed by *vect* and the common locations are to be copied from the index *vect* into the newly generated index *newIndex* for each new possible path passing from an adjacent location. The function *getEndpoint(int, int, int**)* returns the *Integer* value 1 if the locations represented by first two arguments are present in the array and 0 if they are not. It changes the global variable *branchEnd* to the index where the current location coordinate will be inserted and the function *trace()* will be executed on it along with that *newIndex* pointing it for being further copied if any junction comes.

4 Dynamic Memory Based Algorithm (DMA)

4.1 Solved Issues of SMA

Dynamic Memory based Algorithm *(DMA)* uses the *Linked List* because it is a memory efficient data structure. *SMA* was based on the static memory that was allocated before the actual execution of the algorithm. *DMA* shows a significant improvement in the space complexity. In the data structure *pathVector[][][]* shown in Fig. 1, there are unutilized memory blocks which remains unutilized throughout the execution. This issue of wastage of useful memory under *SMA* needs to be solved. *DMA* allocates only necessary chunks of memory while execution for computation. As a specific part of the execution ends, it deallocates the memory used during such execution so that the same memory can be allocated again. The vital benefit of *DMA* is that, the memory parts of all the *pathVector[][][2]* which were used to show whether the location coordinates are saved or not, making a distinction, is not needed in *DMA* because *Linked List* will allocate and append only the necessary memory chunks and when *NULL* comes, it will be the end of the *Linked List*.

4.2 Data Structure, Converted Methods and Results

The entire data structure in *DMA* is dynamically allocated, there is no static allocation. The *Structure* of Language C is used, variables of the *Structure* are passed by reference into the function so that the function can manipulate them. In Fig. 4, *Structure panel* has a link with the *Structure gateway* that contains coordinates and its own pointer, which will point the next structure representing the next location. Here, the map size is variable and the algorithm is executed on different patterns of the map. Coordinates of the map starts from 0 which can be calibrated to 1 as well (Fig. 5).

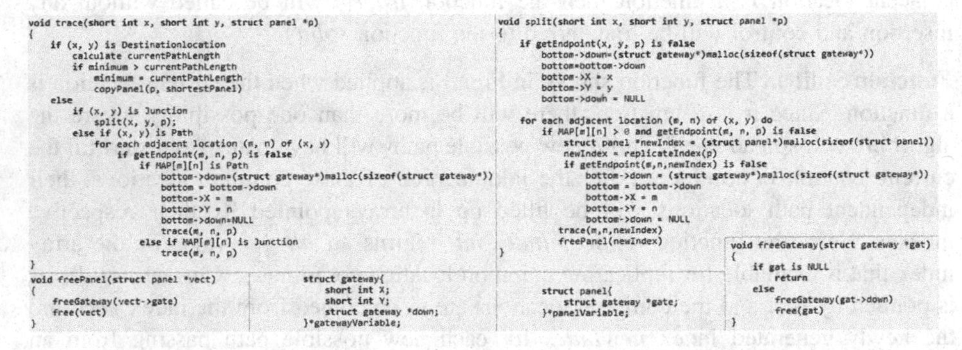

```
void trace(short int x, short int y, struct panel *p)
{
    if (x, y) is Destinationlocation
        calculate currentPathLength
        if minimum > currentPathLength
            minimum = currentPathLength
            copyPanel(p, shortestPanel)
    else
        if (x, y) is Junction
            split(x, y, p);
        else if (x, y) is Path
            for each adjacent location (m, n) of (x, y)
                if getEndpoint(m, n, p) is false
                    if MAP[m][n] is Path
                        bottom->down=(struct gateway*)malloc(sizeof(struct gateway*))
                        bottom = bottom->down
                        bottom->X = m
                        bottom->Y = n
                        bottom->down=NULL
                        trace(m, n, p)
                    else if MAP[m][n] is Junction
                        trace(m, n, p)
}

void split(short int x, short int y, struct panel *p)
{
    if getEndpoint(x, y, p) is false
        bottom->down=(struct gateway*)malloc(sizeof(struct gateway*))
        bottom=bottom->down
        bottom->X = x
        bottom->Y = y
        bottom->down = NULL

    for each adjacent location (m, n) of (x, y) do
        if MAP[m][n] > 0 and getEndpoint(m, n, p) is false
            struct panel *newIndex = (struct panel*)malloc(sizeof(struct panel))
            newIndex = replicateIndex(p)
            if getEndpoint(m,n,newIndex) is false
                bottom->down = (struct gateway*)malloc(sizeof(struct gateway*))
                bottom = bottom->down
                bottom->X = m
                bottom->Y = n
                bottom->down = NULL
            trace(m,n,newIndex)
            freePanel(newIndex)
}

void freePanel(struct panel *vect)
{
    freeGateway(vect->gate)
    free(vect)
}

struct gateway{
    short int X;
    short int Y;
    struct gateway *down;
}*gatewayVariable;

struct panel{
    struct gateway *gate;
}*panelVariable;

void freeGateway(struct gateway *gat)
{
    if gat is NULL
        return
    else
        freeGateway(gat->down)
        free(gat)
}
```

Fig. 4. *Structure gateway* and *panel* (implemented in Language C); modified functions *trace()* and *split()*; *freePanel()* and *freeGateway()* are functions for the memory deallocation.

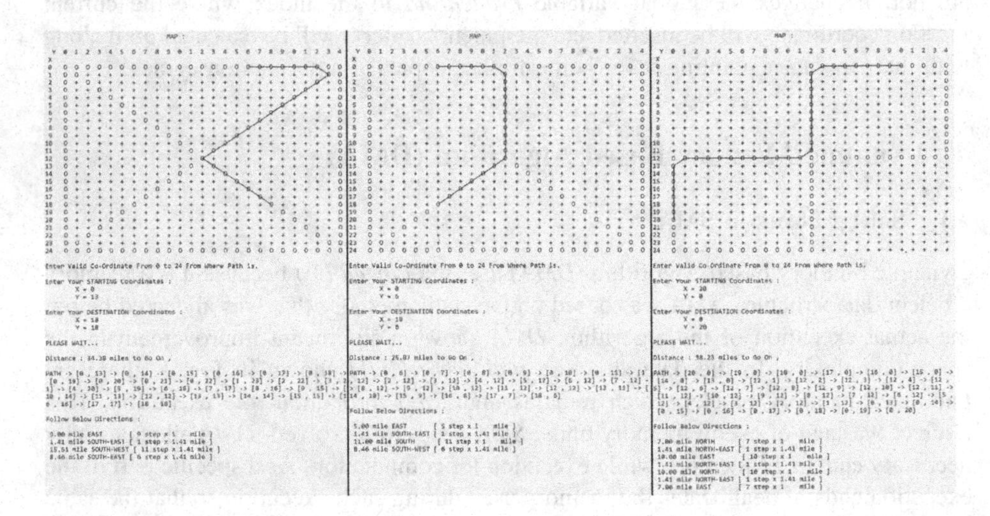

Fig. 5. The shortest path with navigation is generated in (Windows 10) terminal for different patterns of the map. In the map, 'O' indicates presence of the path on that coordinate and '+' represents an empty area.

5 Cluster Based Node Reduction Algorithm (CNRA)

5.1 Solved Issues of SMA and DMA

This approach gives a major breakthrough to *DMA* in dealing with much more complex maps particularly when there are cluster of coordinates close-by, representing the same nearby location. *DMA* is unessentially accurate which is not required in the real-world. By referring [2, 3], multiple interconnected junctions can be represented by one junction representing all of them. Hence, there is only one *split()* instead of multiple splits which implies a reduction in the time and space complexity.

5.2 Algorithm with Auxiliary Functions

Cluster based Node Reduction Algorithm *(CNRA)* creates cluster of multiple junctions connected to each other representing at least 3 locations which are connected to their respective paths. Additional data structures in terms of 3D and 4D arrays are kept as arrays because they are accessed many times while execution. *CNRA* has both static and dynamic allocation schemes. Before the actual execution, the function *analyzeLayer()* is called for each coordinate of the *MAP*. It calls other functions *traceLayer(), exactCenter(), assignCenter()* & *allocateBranch()* once for each cluster sequentially. For each cluster, *traceLayer()* starts from one of its junctions and spreads to each of the junction to calculate how many junctions are residing in the cluster. It sums up each coordinate of the junctions as a weight to find out the center of the cluster. *(centerX, centerY)* is assumed to be the center of the cluster which is further confirmed for being the proper center and if the calculated center is not on the junction of the cluster, then *exactCenter()* assigns it its nearby location i.e. a junction (Fig. 6).

```
void traceLayer(int x, int y, int &sumX, int &sumY, int &count)
{
    mark MAP[x][y]
    sumX = sumX + x + 1
    sumY = sumY + y + 1
    increament count

    for each adjacent location (m, n) of (x, y) do
        if MAP[m][n] is unmarked junction
            traceLayer(m, n, sumX, sumY, count)
}
void assignCenter(int x, int y, int centerX, int centerY, int count)
{
    mark MAP[x][y]
    centerMap[x][y][0] = centerX
    centerMap[x][y][1] = centerY
    centerMap[x][y][2] = count

    for each adjacent location (m, n) of (x, y) do
        if MAP[m][n] is unmarked junction
            assignCenter(m, n, centerX, centerY, count)
}
void getBranch(int x, int y, int &branchX, int &branchY)
{
    for each adjacent location (m, n) of (x, y) do
        if MAP[m][n] is path
            branchX = m
            branchY = n
            return
}
void analyzeMap()
{
    for each coordinate of MAP[i][j] do
        if MAP[i][j] is 2
                analyzeLayer(i, j)
}
```

```
void allocateBranch(int x, int y, int centerX, int centerY, int count)
{
    mark MAP[x][y]
    branchX = branchY = -1

    getBranch(x, y, branchX, branchY)

    linkMap[centerX][centerY][count][0] = branchX
    linkMap[centerX][centerY][count][1] = branchY
    increament count

    for each adjacent location (m, n) of (x, y) do
        if MAP[m][n] is unmarked junction
            allocateBranch(m, n, centerX, centerY, count)
}
void analyzeLayer(int x, int y)
{
    sumX = sumY = count = 0
    traceLayer(x, y, sumX, sumY, count)

    centerX = accuCX = (float)sumX / (float)count - 1
    centerY = accuCY = (float)sumY / (float)count - 1

    if accuCX - (int)accuCX > 0.5
        increament centerX
    if accuCY - (int)accuCY > 0.5
        increament centerY
    if MAP[centerX][centerY] is not junction
        nearestX = accuCX, nearestY = accuCY, leastD = infinity
        exactCenter(x, y, nearestX, nearestY, centerX, centerY, leastD)

    assignCenter(x, y, centerX, centerY, count)
    unmark all marked junction
    allocateBranch(x, y, centerX, centerY, 0)
    unmark all marked junction
}
```

Fig. 6. Definitions and bodies of *CNRA* auxiliary functions.

The function *assignCenter()* traverses through all the junctions, assigns the center coordinate value and the number of junctions to the arrays, which are represented by the junctions of the cluster in *centerMap[][][][]*. Whenever the algorithm approaches any of the junction, it will move to the center for the junction and then it will spread towards each path connected to that cluster. The center of the cluster is associated with the paths connected by the junctions. Function *allocateBranch()* links those paths in 2D array represented by the location of the center for junction on 4D array *linkMap[][][][]*. Function *getBranch()* gives coordinates of path connected with a junction for assigning with *allocateBranch()*.

5.3 Proposed Data Structure

Data structures *centerMap[][][]* and *linkMap[][][][]* are added to *DMA* with their auxiliary functions. The *centerMap[][][]* is a 3D array whereas *linkMap[][][][]* is a 4D array. If (x, y) is a junction and the remaining junctions are connected to it then, for (a, b) representing junctions of the cluster, *centerMap[a][b][3]* will be allocated. Center of cluster (C_x, C_y) is inserted into the *centerMap[a][b][0]* and *centerMap[a][b][1]* respectively, and the number of junctions per cluster is inserted into the *centerMap[a][b][2]*. For *linkMap[][][][]*, the number of junctions $Junction_{total}$ per cluster is calculated and (C_x, C_y) on *linkMap[*C_x*][*C_y*][*$Junction_{total}$*][2]* memory is allocated during the runtime. Coordinates of the junctions will be saved in it, if any junction is surrounded by junctions then (–1, –1) will be inserted. In Fig. 7, *MAP[11][11]* represents map with paths and junctions. There are 5 clusters, four of them have four junctions and one has 5 junctions; e.g. cluster of junctions J = {(4, 0), (5, 0), (6, 0), (5, 1)} have (C_x, C_y) = (5, 0). Now, $\forall$ (a, b) $\in$ J, *centerMap[a][b][0]* = 5, *centerMap[a][b][1]* = 0, *centerMap[a][b][2]* = 4 because there are four junctions in the cluster. Those junctions are connected with (3, 0), *NULL*, (7, 0), (5, 2) respectively. These coordinates represents the paths (branches) connected to the cluster and assigned to the *linkMap[*C_x*= 5][*C_y*= 0][i][2]* one by one. For example, path (7, 0) is saved in *linkMap[5][0][0][0]* and *linkMap[5][0][0][1]* respectively. When algorithm executes *split()* from any junction, it will investigate *linkMap[*C_x*][*C_y*][][]* and sequentially spreads through each paths i.e. if control is on coordinate (6, 0) then it will spread through each coordinate *(linkMap[*C_x*][*C_y*][i][0], linkMap[*C_x*][*C_y*][i][1])* where $0 \leq i \leq 3$. If it finds (–1, –1) then it will not proceed due to absence of the path.

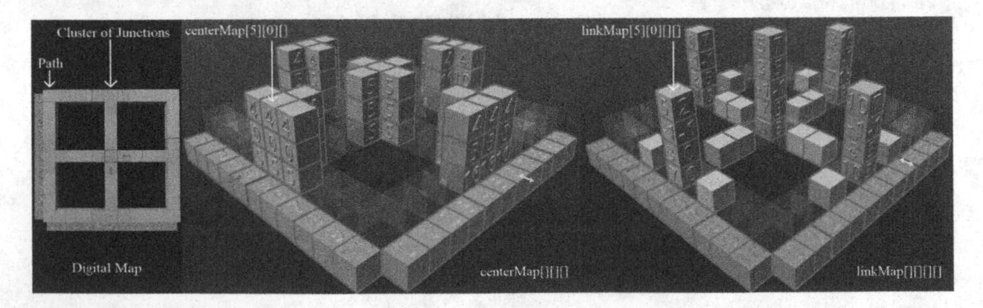

Fig. 7. 3D representation of *centerMap[][][]* and *linkMap[][][][]* (using Microsoft 3D Builder).

6 Connected Cluster Based Traversing Algorithm (CCTA)

6.1 Traversal Overhead in CNRA

CNRA is based on digital map traversal, where the algorithm finds the path towards the adjacent location. If there is a path instead of a junction then the algorithm will simply traverse through all the continuous locations representing the same path. On every traversal of each path from their adjacent junctions, many locations are traversed invariably and this is the overhead. The algorithm should therefore, jump onto one junction from another directly without traversing intermediate nodes.

6.2 Connected Clusters Approach and Its Significance

Connecting clusters over their path removes overhead produced by traversing the entire path. If each cluster knows its adjacent clusters connected to it then, each cluster will be identified as a node and will be traversed directly to its adjacent clusters. Functionalities of *DMA* and *CNRA* are preserved in Connected Cluster based Traversing Algorithm *(CCTA)*, only intermediate approach is modified. For applying this approach, the data structure of *CCTA* is modified and the required functions are added. Here, data structure *linkMap* is extended to its 4th dimension to length 5 i.e. *linkMap[][][][5]*. First two values represent branch coordinates, 3rd and 4th values represent coordinates of adjacent cluster connected to those branch coordinates and 5th value represents the length of the path between the current cluster and the adjacent cluster. These values are calculated before the actual execution of *CCTA*.

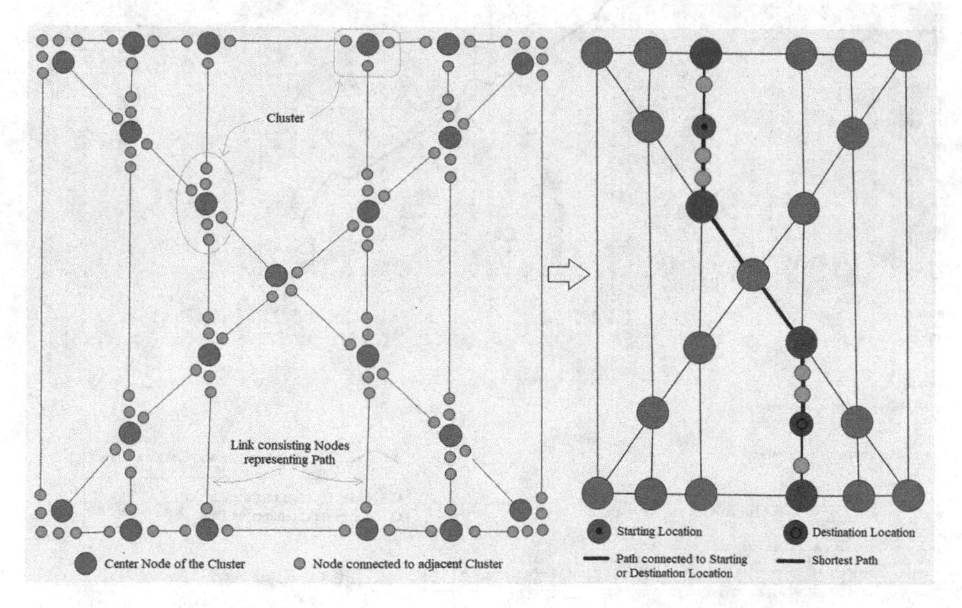

Fig. 8. Transformation of a digital map (Fig. 9) into the cluster based map and further conversion into the connected cluster (node) based network. (Color figure online)

The shortest path calculated by *CCTA* is illustrated in Fig. 9 on 50 × 50 digital map. The shortest path from (10, 16) to (40, 34) is generated and drawn in Fig. 9. *CCTA* traces one by one location when it is on the paths consisting of a starting location and a destination location i.e. paths (2, 6) to (13, 6) and (37, 34) to (47, 34) because there will be no option of the jump. Initially, starting location acts like a junction (if it is on the path) to traverse towards two ways. Once it reaches a cluster and if the destination location is not on any of the paths connected to that cluster, then it will jump to an adjacent node without traversing nodes of intermediate paths, else it will traverse all nodes of the only path that leads to the destination location. Cluster represented by dark-colored nodes are connected to either starting or destination location in Fig. 8, remaining clusters are connected to each other so that the traversal can jump directly while execution. Though *SMA* can be integrated with *CCTA*, the methodology for saving traversed locations is the same as that of the *CNRA* i.e. using *Linked List* so that, if in case the map is much complex, the segmentation fault does not occur. The map is reduced for faster execution and the original data is used.

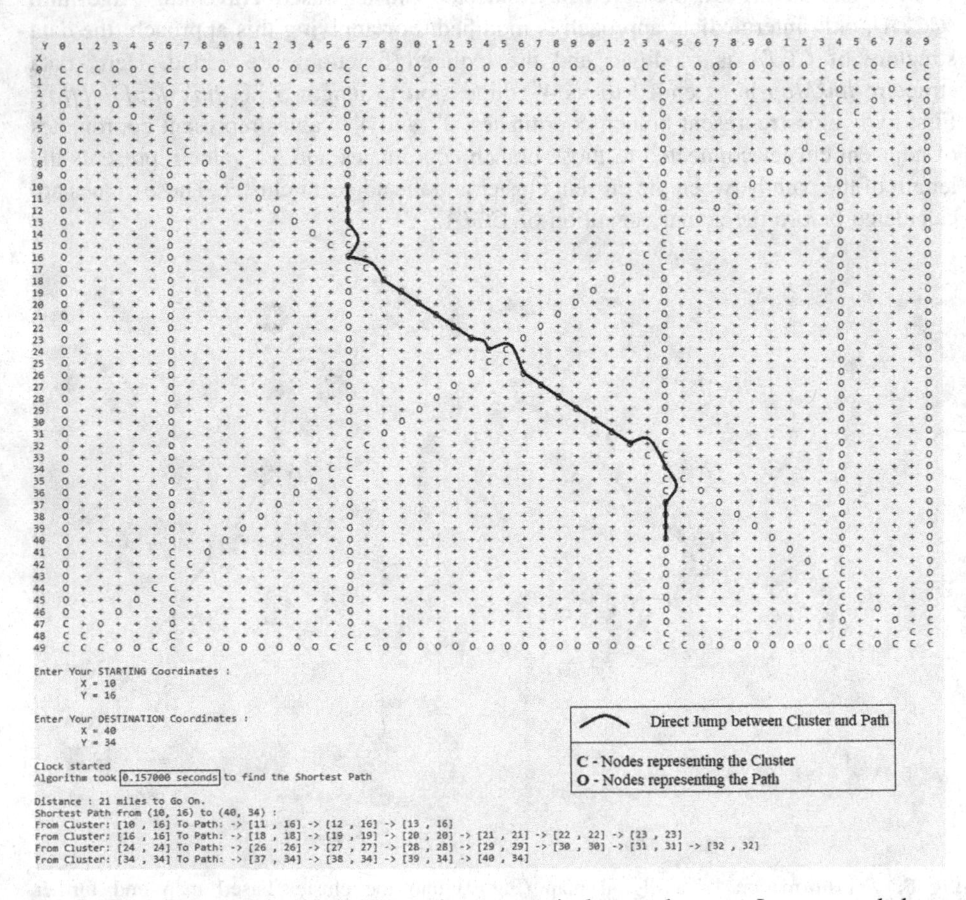

Fig. 9. The shortest path with *CCTA* on the comparatively complex map. Its connected cluster based node representation is shown in Fig. 8.

7 Comparison Among Approaches

An overview of the execution time and memory consumption are illustrated in Figs. 10, 11 and 12. Memory consumption of *SMA* is 19 GB.

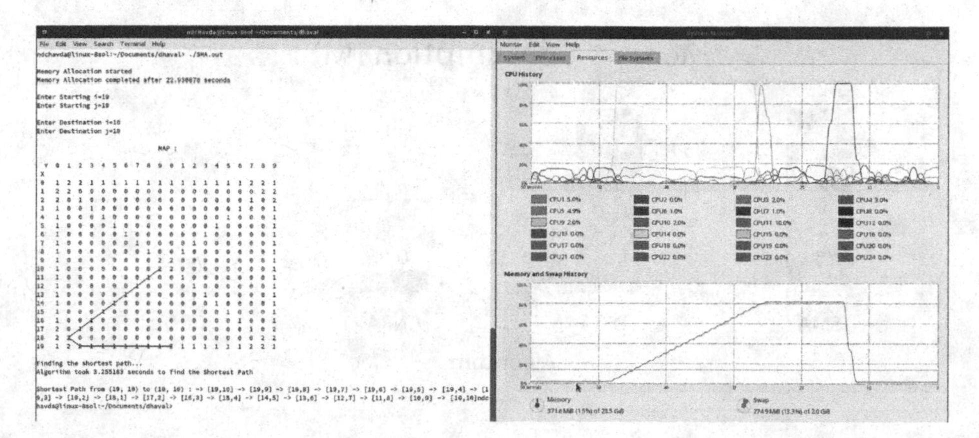

Fig. 10. *SMA* execution (Workstation, OpenSUSE, Intel Xeon, 24 Core CPU, 24 GB RAM).

```
                         MAP :

 Y  0 1 2 3 4 5 6 7 8 9 0 1 2 3 4 5 6 7 8 9
 X
 0  C C C O O O O O O O O O O O O O O C C C
 1  C C + + + + + + + + + + + + + + + + C C
 2  C + O + + + + + + + + + + + + + + O + C
 3  O + + O + + + + + + + + + + + + O + + O
 4  O + + + O + + + + + + + + + + O + + + O
 5  O + + + + O + + + + + + + + O + + + + O
 6  O + + + + + O + + + + + + O + + + + + O
 7  O + + + + + + O + + + + O + + + + + + O
 8  O + + + + + + + O + + O + + + + + + + O
 9  O + + + + + + + + C C + + + + + + + + O
10  O + + + + + + + + C C + + + + + + + + O
11  O + + + + + + + O + + O + + + + + + + O
12  O + + + + + + O + + + + O + + + + + + O
13  O + + + + + O + + + + + + O + + + + + O
14  O + + + + O + + + + + + + + O + + + + O
15  O + + + O + + + + + + + + + + O + + + O
16  O + + O + + + + + + + + + + + + O + + O
17  C + O + + + + + + + + + + + + + + O + C
18  C C + + + + + + + + + + + + + + + + C C
19  C C C O O O O O O O O O O O O O O C C C
```

```
Enter Valid Co-Ordinate From 0 to 19 From Where Path is.

Enter Your STARTING Coordinates :
          X = 19
          Y = 10

Enter Your DESTINATION Coordinates :
          X = 11
          Y = 8

Clock started
Algorithm took 0.015000 seconds to find the Shortest Path

Shortest Path from (19, 10) to (11, 8) :-> [19 , 10] -> [19 , 9]
 -> [19 , 8] -> [19 , 7] -> [19 , 6] -> [19 , 5] -> [19 , 4] ->
[19 , 3] -> [18 , 1] -> [17 , 2] -> [16 , 3] -> [15 , 4] -> [14
, 5] -> [13 , 6] -> [12 , 7] -> [11 , 8]
```

Fig. 11. *CNRA* execution time (PC, Windows 10, Intel Core i7, Quad Core CPU, 16 GB RAM).

```
 Y  0 1 2 3 4 5 6 7 8 9 0 1 2 3 4 5 6 7 8 9
 X
 0  C C C O O O O O O O O O O O O O O C C C
 1  C C + + + + + + + + + + + + + + + + C C
 2  C + O + + + + + + + + + + + + + + O + C
 3  O + + O + + + + + + + + + + + + O + + O
 4  O + + + O + + + + + + + + + + O + + + O
 5  O + + + + O + + + + + + + + O + + + + O
 6  O + + + + + O + + + + + + O + + + + + O
 7  O + + + + + + O + + + + O + + + + + + O
 8  O + + + + + + + O + + O + + + + + + + O
 9  O + + + + + + + + C C + + + + + + + + O
10  O + + + + + + + + C C + + + + + + + + O
11  O + + + + + + + O + + O + + + + + + + O
12  O + + + + + + O + + + + O + + + + + + O
13  O + + + + + O + + + + + + O + + + + + O
14  O + + + + O + + + + + + + + O + + + + O
15  O + + + O + + + + + + + + + + O + + + O
16  O + + O + + + + + + + + + + + + O + + O
17  C + O + + + + + + + + + + + + + + O + C
18  C C + + + + + + + + + + + + + + + + C C
19  C C C O O O O O O O O O O O O O O C C C
```

```
Enter Your STARTING Coordinates :
          X = 12
          Y = 7

Enter Your DESTINATION Coordinates :
          X = 19
          Y = 10

Clock started
Algorithm took 0.000000 seconds to find the Shortest Path

Distance : 14.05 miles to Go On.
Shortest Path from (12, 7) to (19, 10) :
From Cluster: [12 , 7] To Path: -> [13 , 6] -> [14 , 5] -> [15 ,
 4] -> [16 , 3] -> [17 , 2]
From Cluster: [18 , 1] To Path: -> [19 , 3] -> [19 , 4] -> [19 ,
 5] -> [19 , 6] -> [19 , 7] -> [19 , 8] -> [19 , 9] -> [19 , 10]
```

– Execution time of *CCTA* is in microsecond (calculated on complex map) so that 0.000000 is displayed by Language C function - clock() as per its accuracy

Fig. 12. *CCTA* execution time (PC, Windows 10, Intel Core i7, Quad Core CPU, 16 GB RAM).

CNRA decreases the execution time to the milliseconds, and *CCTA* further decreases it to the microseconds, memory usage being in KB in both the approaches (Figs. 13, 14 and Table 1).

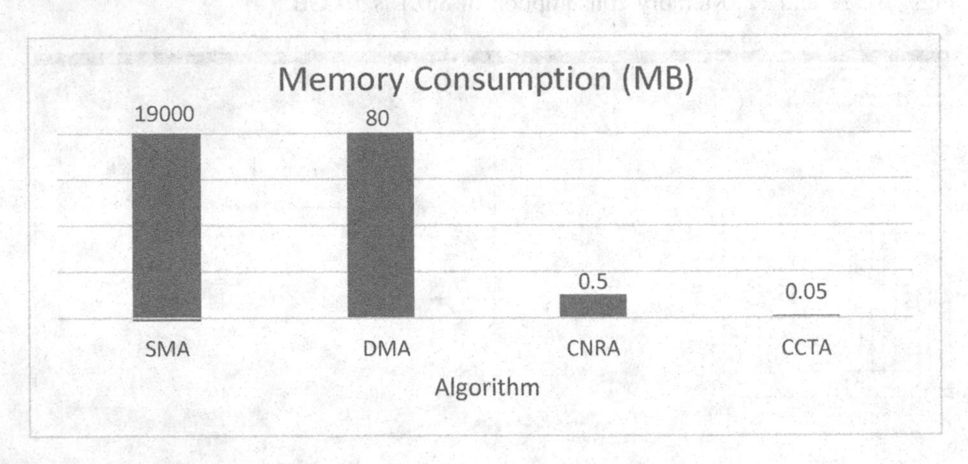

Fig. 13. Memory consumption for above approaches.

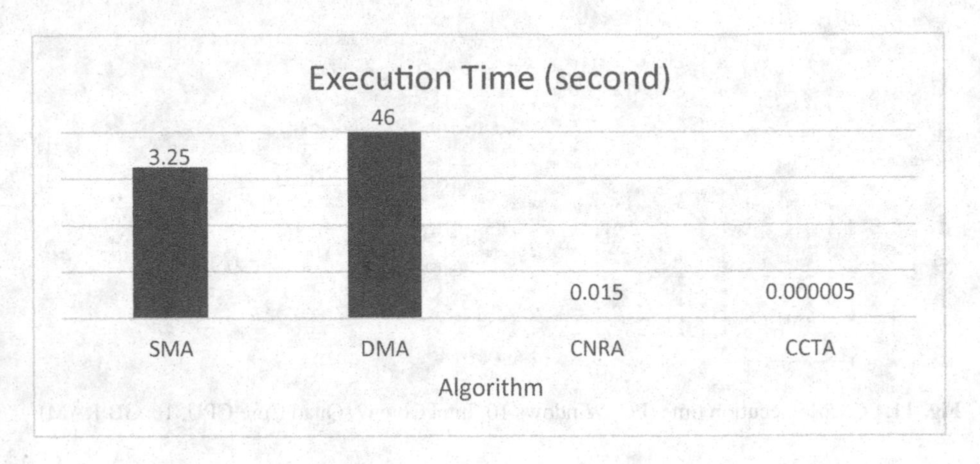

Fig. 14. Execution time for above approaches.

Table 1. Performance comparison among approaches.

	Execution time	Memory usage	Machine
SMA	3.25 s	19 GB	Workstation, Intel Xeon, 24 Core CPU, 24 GB RAM
DMA	46 s	80 MB	PC, Intel Core i7, Quad Core CPU, 16 GB RAM, Clock speed 3.4 GHz
CNRA	15 ms	500 KB	Same as *DMA*
CCTA	5 μs	50 KB	Same as *CNRA*

8 Conclusion

Thus, by the aforementioned analysis and the implementation of the algorithms, it is observed that *SMA* requires plenty of memory because of its static allocation. *DMA* on the contrary requires less memory but needs more execution time because it is entirely based on dynamic memory allocation – algorithm allocates and deallocates memory for further use which takes the CPU time (clock cycles). *CNRA* solves problems of both *SMA* and *DMA* by implementing auxiliary methods which simplify the execution. Clustering of junctions decreases the number of effective nodes and extends feasibility for much complex maps. *CCTA* bypasses the traversal of the path directly to the cluster so that it removes unnecessary intermediate traversal overhead. In doing so, it gives the same result as *CNRA*. In addition, it improves the execution time and memory consumption.

This paper illustrates how the digital map is represented and processed in its nodal form. *CCTA* can be elaborated like this – "Transform the given data into its compact form, process its compact form, generate an intermediate result and map it to the original data for an accurate result" i.e. transform the digital map into its cluster based node representation, process and map it to the digital map for obtaining the real-world shortest path.

These approaches can further be extended by compressing the nodal map of *CCTA* by some layers for simplifying it. Each layer can be mapped in a sequence. After having the shortest path by processing the compact map of the bottommost layer, its mapping with an above layered map can be used for obtaining an equivalent shortest path which is much apparent. Repeating the same procedure till the digital map i.e. till the topmost layer will give the real-world shortest path. Moreover, if the algorithm is modified to make it follow the direction towards the destination, it will eliminate the false positive directions further. In case, the algorithm is unable to find the destination, other possibilities of directions can be listed into the priority queue that depends upon the gradient of the current location to the destination location. The direction based approach is fast and intelligent while the earlier recursive approach is accurate, hence they exhibits the time-accuracy tradeoff.

References

1. Giorgio Gallo, F., Stefano Pallottino, S.: Shortest path algorithms. Ann. Oper. Res. **13**(1), 1–79 (1988). https://doi.org/10.1007/BF02288320
2. Krishna, P., Vaidya, N.H., Chatterjee, M., Pradhan, D.K.: A cluster-based approach for routing in dynamic networks. ACM SIGCOMM Comput. Commun. Rev. **27**(2), 49–64 (1997). https://doi.org/10.1145/263876.263885
3. Zhang, D., Yang, D., Wang, Y., Tan, K. L., Cao, J., Shen, H.T.: Distributed shortest path query processing on dynamic road networks. VLDB J. **26**(3), 399–419 (2017). https://doi.org/10.1007/s00778-017-0457-6
4. Kadia, D.: Binary search optimization: implementation and amortized analysis for splitting the binary tree. Int. J. Comput. Sci. Inf. Secur. **15**(3), 338–341 (2017)

Learning Cuckoo Search Strategy for t-way Test Generation

Abdullah B. Nasser, Abdulrahman Alsewari, and Kamal Z. Zamli(✉)

Faculty of Computer Systems and Software Engineering,
Universiti Malaysia Pahang, 26300 Kuantan, Pahang, Malaysia
Abdullahnasser83@gmail.com, alsewari@gmail.com,
kamalz@ump.edu.my

Abstract. The performance of meta-heuristic algorithms highly depends on their exploitation and exploration techniques. In the past 30 years, many meta-heuristic algorithms have been developed which adopts different exploitation and exploration techniques. Several studies reported that the hybrid of meta-heuristics algorithms often perform better than its corresponding original algorithm. This paper presents a new hybrid algorithm; called Learning Cuckoo Search (LCS) strategy based on the integration student phase from Teaching Learning based Optimization (TLBO) Algorithm. To evaluate the developed algorithm, we use the problem of t-way test generation as our case study. The experiment results show that LCS has better performance as compared as to the original Cuckoo Search as well many other existing strategies.

1 Introduction

The performance of meta-heuristic algorithms highly depends on their search technique capabilities. Often, the performance of meta-heuristic algorithms depends on their exploitation and exploration strategy. Exploitation explores the promising regions in the hope to find better solutions while the exploration ensures that all regions of the search space have been visited. Usually, a good balance between intensive and efficient exploration plays an important part in the performance of meta-heuristic algorithms [1].

Many meta-heuristic algorithms have been developed in the past 30 years such as Tabu search (TS) [2], Simulated Annealing (SA) [3], Genetic Algorithm (GA) [4], Particle Swarm Optimization (PSO) [5], Ant Colony Optimization (ACO) [6], Differential Evolution (DE) [7], Harmony Search HS [8], Flower Pollination Algorithm (FPA) [9], Sine Cosine Algorithm (SCA) [10], Bat Algorithm (BA) [11], Cuckoo Search (CS) [12] Teaching–Learning-Based Optimization Algorithm (TLBO) and Firefly Algorithm (FA) [13]. Meta-heuristic algorithms have been successfully used for solving a wide range of software engineering problems on software engineering management, requirements engineering, design, testing, and refactoring.

Concerning exploitation and exploration techniques, GA, for instance, adopts selection, and crossover, mutation operations. PSO adopts as two search approach: search around overall best (gbest) and search around personal best (pbest). CS depends on Levy flight and elitism technique. TLBO divides the search into teacher and learner phases.

R. Sharma et al. (Eds.): ICAN 2017, CCIS 805, pp. 97–110, 2018.
https://doi.org/10.1007/978-981-13-0755-3_8

Each algorithm has its strengths and limitations as there is no single algorithm that is superior for all optimization problems. For these reasons, the search for a new algorithm is justified by developing a new meta-heuristics algorithm. Experiments studies reported that the hybrid of meta-heuristics algorithms often perform better than its corresponding origin algorithm [14]. For example, He et al. demonstrated a hybridized Variable Neighbourhood search with TS to minimize the discrete time/cost trade-off problem [15]. Yildiz presented a new hybrid algorithm based on Hill Climbing local search and Artificial Immune Algorithm for solving general optimization problem [16]. Wang et al. adopted SA and GA algorithms to optimize the cutting conditions in plain milling [17].

Building from the aforementioned prospect, the main focus of this work to present a new hybrid algorithm, called Learning Cuckoo Search strategy (LCS), based on the integration of Cuckoo search with the student phase from Teaching Learning based Optimization algorithm (TLBO). Our hybridization approach is unique from existing hybridizations of CS as we use the peer learning phase operator during the elitism phase of the Cuckoo search algorithm. As a case study, we adopt LCS for the t-way test generation problem. In a nutshell, t-way test generation problem is a sampling technique to select a sub-set of test cases that can be used to test overall system such that every t combination of input values (where t refer to interaction strength) is covered at least one time [18].

To this end, much recent works on t-way strategies are focusing using a single meta-heuristic algorithm (such as TS, SA, GA, ACA, PSO, and HS, to name a few [18–21]). Although useful, existing work has not sufficiently dealt with hybrid meta-heuristic algorithm (i.e. combinations of two or more meta-heuristic algorithms) as the backbone for t-way strategies. Taking this challenge has led us toward the current work.

The organization of this paper is as follows. In Sect. 2, an overview on how t-way testing works is provided. Section 3 reviews existing t-way strategies. Section 4 describes the proposed strategy, while Sect. 5 highlights and discusses the results. Lastly, Sect. 6 gives the conclusion and future work.

2 Problem Definition of t-way Testing

To illustrate the concept of t-way testing, consider a Car Ordering System (COS) example. The system allows for buying and selling cars in Malaysia. The system allows two types of booking; online booking or in store during Opening hours or closing hours. Here, this system consists of five inputs (i.e. Order category, Location, Car brand, Order type and Order time), three parameters with two values, one parameter with nine values, and one parameter with one value. The system can be summarized in Table 1.

In COS, there are 72 combinations need to be tested for ideal testing (i.e. exhaustively testing which considers all interaction strength, t = 5). By considering two-way interaction, the reduction of test suite size can be achieved 72 to 9 test cases as shown in Table 2.

Table 1. Car ordering system parameters

Order category	Location	Car brand	Order type	Order time
Buy	Kuala Lumpur	Perodua	E-Booking	Opening hours
Selling	Penang	Toyota	In store	Closing hours
	Johor Bahru	Proton		

Table 2. Two-way test suite for car ordering system

No	Order category	Location	Car brand	Order type	Order time
1	Buy	Kuala Lumpur	Perodua	E-Booking	Opening hours
2	Selling	Penang	Perodua	In store	Closing hours
3	Buy	Johor Bahru	Toyota	In store	Opening hours
4	Selling	Johor Bahru	Proton	E-Booking	Closing hours
5	Buy	Penang	Toyota	E-Booking	Closing hours
6	Selling	Kuala Lumpur	Proton	In store	Opening hours
7	Buy	Penang	Proton	E-Booking	Opening hours
8	Selling	Kuala Lumpur	Toyota	In store	Closing hours
9	Selling	Johor Bahru	Perodua	E-Booking	Closing hours

The same generalization can be done for 3-way and so forth. It is up to the creativity and the knowledge of the test engineers to decide on the right t value based on the testing requirement at hand. Researchers often advocate the best range of values for t is from 2 to 6.

3 Related Work

Generally speaking, there are two main domains of the existing works on t-way testing: algebraic and computational methods [22]. Algebraic methods construct based on lightweight mathematical functions without enumerate any combinations. Strategies of this approach (e.g. Combinatorial Test Services (CTS) strategy and TConfig [23]) are restricted for small configurations ($t \leq 3$).

Computational methods offer flexibility to address large configuration. As the name suggests, computational approaches use pure computation strategies or meta-heuristic algorithms to construct the test cases. Strategies adopting computational methods can be categorized into two categories: one-test-at-a-time (OTAT) and one-parameter-at-a-time (OPAT) strategies. Computational methods mainly based on generating all possible combinations (i.e. interaction elements) according to system's configurations.

OPAT strategies start by constructing a completed test suite for the smallest interaction parameters, then in every iteration one parameter (i.e. one column) are added until all the parameters are covered. In-parameter-order (IPO) [24] is the pioneer work in this respect. Many improvements to the basis of IPO strategy have been developed such as that of IPOG-D [25], IPOG [26], IPOF and IPAD2 [27] to obtain the smallest test sizes and fast execution times.

Concerning OTAT strategies, a complete test case is constructed per iteration that covers the maximum number of uncovered interaction elements. The same procedure is repeated until all interaction elements are covered. In the literature, numerous tools and strategies have developed based on OTAT approach such as Automatic Efficient Test Generator (AETG) [28], Classification-Tree Editor eXtended Logics (CTE-XL) [29], Pairwise Independent Combinatorial Testing (PICT) [30], Deterministic Density Algorithm (DDA) [31, 32], Test Vector Generator (TVG) [33], GTWay [34], Jenny [35], and WHITCH [36].

Recently, meta-heuristic algorithms have been adopted as the backbone for t-way test suite generation. In general, meta-heuristic-based strategies start with a random set of solutions. These solutions are subjected to a series of search operation in an attempt to improve them. During each iteration, the best candidate solution is selected and added to the final test suite. In the literature, many meta-heuristic algorithms have been successfully applied for t-way testing such as TS [37], SA [38], GA [39], ACA [39], PSO [22, 40], HS [18], and CS [41].

Stardom [20] presented a description of adopting SA, GA and TS algorithms for two-way testing. SA [38] is a single-based physical Algorithm, inspired from the physical annealing process. The algorithm starts searching from one position and then employ neighborhood search in a local region in attempt to find better solution. SA allows moving to poor solution with acceptance probability to avoid stuck in a local minimum solution. GA is an early algorithm for adopting a population-based algorithm in t-way testing. it starts finding optimal test case from many positions and then repeated apply selection, crossover, and mutation operations in order to mimic natural selection of biological evolution. Similar to SA, TS accept a worse move if no improving move is available. TS utilizes memory structures (termed Tabu list) for guiding the search process (i.e. to remember the visited solutions).

Later on, Cohen extended SA to support 3-way interaction testing [42], and Shiba et al. extended GA and ACO to support 3-way interaction testing [43]. ACO mimics the behavior of colonies of ants for finding food paths. Here, each test case represents the quality of the paths to the food and the food represents the value of the parameter. ACA use trails of a chemical substance, called as pheromone which reinforce over time. Pheromone enables other ants to find short paths of the food source. Comparative experiments between SA, TS and GA conducted by Colbourn et al. demonstrate that SA performs better than TS and GA [38].

Chen et al. [44] implemented PSO algorithms for 2-way testing and [21] for t-way testing. The algorithm mimics the swarm behavior of bird and fish swarm in searching food. Based on simple formulae, the population (i.e. called a swarm) moves in the search space, guided by global best and personal best in attempt to find better solution. Recently, [18] adopted HS for design and implementation a new t-way strategy called Harmony Search Strategy (HSS). HS is inspired by the behavior of musicians, to produce a new musical tone. The strategy support high interaction strengths (i.e. $t \geq 15$).

Nasser et al. [45] have implemented the Flower pollination based strategy (FPA) for generating *t*-way test generation. FPA is also used for generating sequence *t*-way test suite [46]. Inspired by the pollination behavior of flowering plants, FPA can be represented as two steps (i.e. Global Pollination and Local Pollination), controlled by probability parameter. Global Pollination step exploits lévy flight to transfer the pollen to another flower while local pollination transfers the pollen to female part within the same flower. In similar work, Alsariera adopted Bat Algorithm for *t*-way testing [47, 48]. The algorithm is inspired by the hunting behavior of Microbats which are able to find its prey in complete darkness.

Recently, Zamli et al. [49] proposed a new hyper-heuristic based strategy called High Level Hyper-Heuristic (HHH). In HHH, Tabu search algorithm serves as the master algorithm (i.e. High level) to control other four low level algorithms(LLH); Particle Swarm Optimization, Teaching Learning based Optimization, Cuckoo Search Algorithm and Global Neighborhood Algorithm. To ensure high performance, HHH defines a new acceptance mechanism for the selection LLH algorithm, relies on three operations (i.e. diversification, intensification and improvement). Further experiments have been done in [50] with new acceptance mechanism based on fuzzy inference system and new LLH operations (i.e. GA's crossover search operator of, TLBO's learning search operator, FPA's global Pollination, and Jaya algorithm's search operator).

Building from earlier approach, Zamli et al. presented [51] a new strategy, called Adaptive Teaching Learning-Based Optimization (ALTBO). ATLBO improves the performance of standard TLBO resulting from a good balance between intensification and diversification through the adoption of fuzzy inference rules.

4 Proposed Strategy

This section describes the Learning Cuckoo Search Strategy (LCS) strategy based on the hybridization of Cuckoo Search with the student phase of Teaching Learning based Optimization algorithm.

4.1 Cuckoo Search

CS is one of the latest nature inspired algorithms inspired from brood parasitic behavior of Ani and Guira cuckoos [12]. Cuckoo has an aggressive reproduction strategy in that they lay their eggs in the nests of other host birds. In order to increase the hatching probability of their own eggs, they often remove the eggs of the host bird.

Concerning its corresponding algorithm, CS initially generates a population of nests randomly. The nest is updated using Levy Flight, based on the following equation:

$$x_i^{(t+1)} = x_i^{(t)} + \alpha \oplus L\acute{e}vy(\lambda) \quad (1)$$

To mimic the removal of host eggs, CS proposes elitism technique which is controlled by the probability (*pa*). Figure 1 summarizes the complete CS algorithm.

Algorithm 1: Cuckoo Search Algorithm

1: Objective function f(x), x = (x1, ..., xd) ;
2: Initial a population of n host nests x_i *(i = 1, 2, ..., n);*
3: **While** (*t* <*MaxGeneration*) or (stop criterion)
4: Get a cuckoo (say *i*) randomly by *Lévyflights*;
5: Evaluate its quality/fitness *Fi;*
6: Choose a nest among n (say *j*) randomly;
7: **IF (Fi > Fj)**
8: Replace j by the new solution;
9: **End if**
10: Abandon a fraction (*pa*) of worse nests and build
11: new ones.
12: Keep the best solutions (or nests with quality solutions);
13: Rank the solutions and find the current best;
14: **End while**
15: Postprocess results and visualization;
16: End-Procedure

Fig. 1. Cuckoo search algorithm [12]

4.2 Teaching-Learning Based Optimization

Teaching-Learning Based Optimization (TLBO) algorithm is proposed by Rao, et al. [52]. The algorithm inspired from the classroom setting between a teacher and learners. In general, TLBO can view as two phases: Teacher Phase and Learner Phase. Here, the student can learn from the teacher or from his/her partner. In TLBO, the population consider as two groups teachers and learners. At each iteration, TLBO undergoes the two phases sequentially.

Teacher Phase attempts to improve individual solution x_i, by moving their position toward their teacher (Eq. 2):

$$x_i^{(t+1)} = x_i^{(t)} + r\left(x_{teacher} - T_f * x_{mean}\right) \tag{2}$$

Learner Phase represents local search's part in TLBO. This phase attempts to explore the solution around each individual. Each individual x_i improves its knowledge by interacting with its random peer x_i then move its position to new learner's position. The individual moves to learner, if only there is improvement (Eq. 3), otherwise move toward Eq. 4:

$$x_i^{(t+1)} = x_i^{(t)} + r\left(x_j - x_i\right) \tag{3}$$

$$x_i^{(t+1)} = x_i^{(t)} + r\left(x_i - x_j\right) \tag{4}$$

4.3 Learning Cuckoo Search

LCS is a composition of two main steps: generating interaction elements which represent the search space and find optimal t-way test suite using meta-heuristic algorithm.

As mentioned previously, the core parts of the CS algorithm are generating new solutions using Levy Flight and replacing the worst nests by new nest using elitism technique. Although CS has good capability for exploration the search space efficiently, many studies have shown CS has local conversion issues as the algorithm can easily be trapped in local minima [53–55]. Addressing this issue, the proposed strategy integrates CS with learner phase from TLBO. Original CS replaces the worst by new nest generated randomly. Our proposed strategy generates the new nests as part of elitism based on learner phase of TLBO. Figure 2 illustrates the proposed Learning Cuckoo Search Algorithm (LCS).

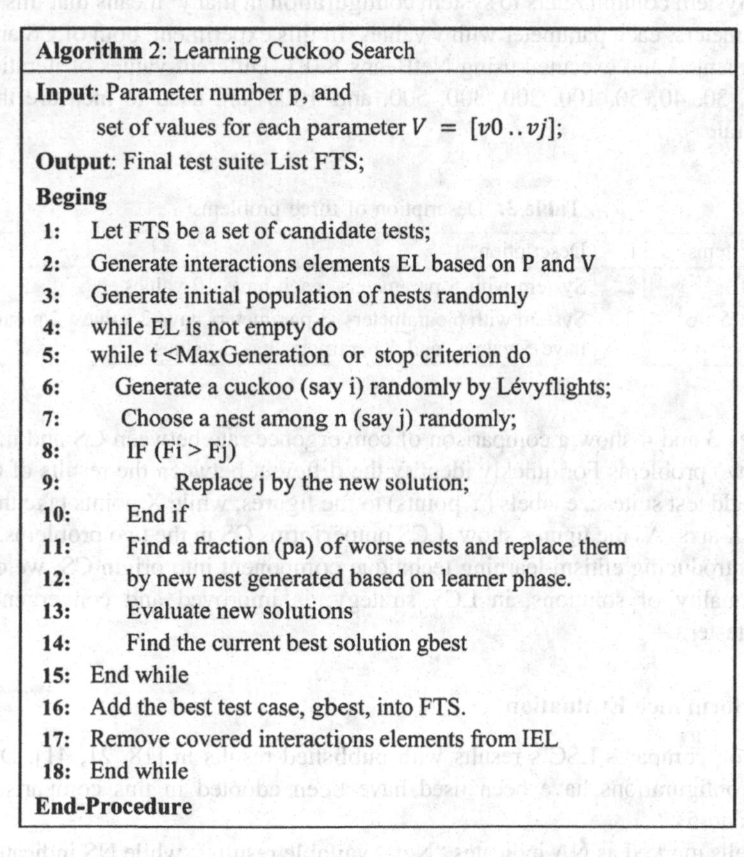

Algorithm 2: Learning Cuckoo Search

Input: Parameter number p, and
set of values for each parameter $V = [v0\,..\,vj]$;
Output: Final test suite List FTS;
Beging
1: Let FTS be a set of candidate tests;
2: Generate interactions elements EL based on P and V
3: Generate initial population of nests randomly
4: while EL is not empty do
5: while t <MaxGeneration or stop criterion do
6: Generate a cuckoo (say i) randomly by Lévyflights;
7: Choose a nest among n (say j) randomly;
8: IF (Fi > Fj)
9: Replace j by the new solution;
10: End if
11: Find a fraction (pa) of worse nests and replace them
12: by new nest generated based on learner phase.
13: Evaluate new solutions
14: Find the current best solution gbest
15: End while
16: Add the best test case, gbest, into FTS.
17: Remove covered interactions elements from IEL
18: End while
End-Procedure

Fig. 2. Learning cuckoo search strategy

5 Results

In order to evaluate the performance of LCS, first, we compare the convergence rate of LCS with its counterparts CS, and then LCS is compared against other existing strategies including CS. The parameters of LCS are set at $Pa = 0.8$, population size = 30 and the maximum number of improvements = 300.

5.1 Convergence Rate Analysis

In order to evaluate the performance of the proposed strategy against the original CS, the convergence rate for different systems is studied. Here, convergence rate is used to measure how fast meta-heuristic algorithms converge (i.e. covering all the required interactions) per generation. We apply CS and LCS on two systems as shown in Table 3. System column refers to system configuration in that y^x means that this system has x parameters, each parameter with y values. In this experiment, both of CS and LCS are implemented and executed using NetBeans 8.0.1. Different values of iteration (i.e. 5, 10, 20, 30, 40, 50, 100, 200, 300, 500, and 1000) are used to measure the convergence rate.

Table 3. Description of three problems

No.	Systems	t	Description
1	10^5	2	System with 5 parameters, each have 10 values
3	$7^3, 5^2, 3^1$	3	System with 6 parameters, 3 parameters have 2 values, 2 parameters have 5 values, and 1 parameter has 3 values

Figures 3 and 4 show a comparison of convergence rate between CS and modified LCS for two problems For quickly identify the different between the results of CS and LCS, we add test suite size labels (Y points) to the figures, while X points take the same values of X aixs. As the figures show, LCS outperforms CS in the two problems. In this case, by introducing elitism-learning technique component into origin CS, we observe that the quality of solutions, in LCS strategy, is improved and convergence rate becomes faster.

5.2 Performance Evaluation

This section compares LSC's results with published results in [18, 21, 41]. Different systems configurations have been used have been adopted in this comparison (see Tables 4 and 5).

The cells marked as NA indicates "Not Available results", while NS indicates "Not Support". Tables 4 and 5 present a comparison of LCS against the existing strategies. The cells marked with bold font present the optimal value achieved by the strategy on the corresponding column. Most of existing strategies, in Table 3, support only t ≤ 3

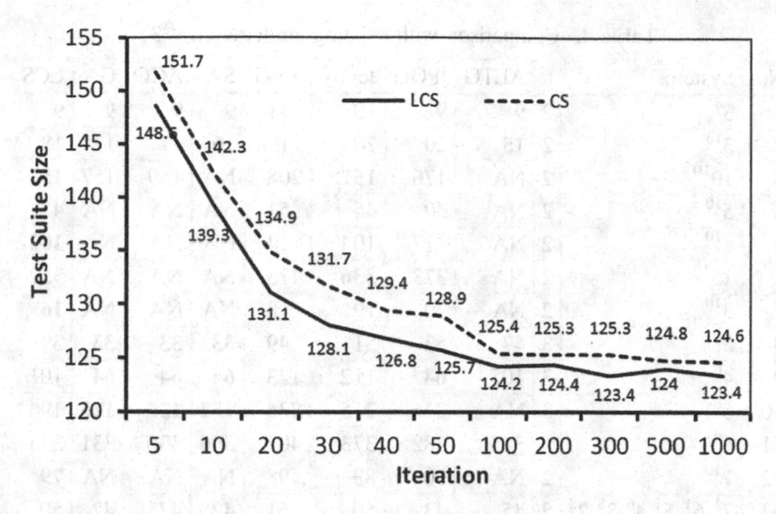

Fig. 3. Convergence rate of CS and LCS for 10^5 with $t = 2$

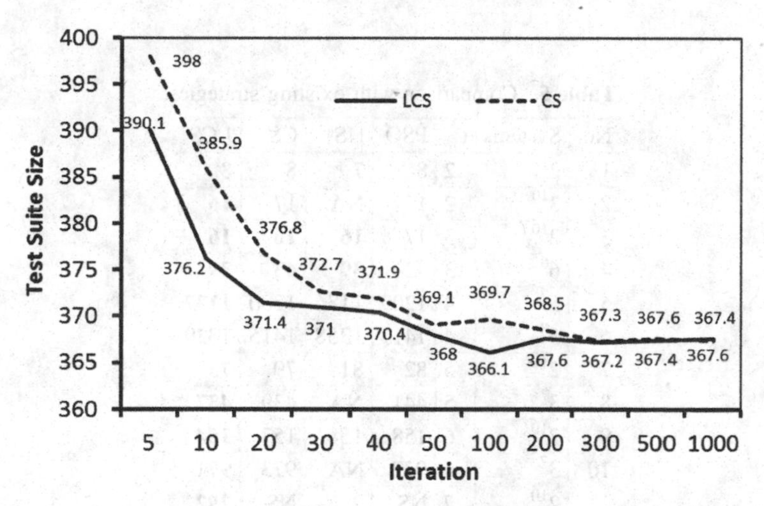

Fig. 4. Convergence rate of CS and LCS for $7^3, 5^2, 3^1$ with $t = 3$

such as GA, SA, and ACO, while Table 4 includes strategies support higher interaction strength $(t > 3)$ such as PSO, HS, and CS. Results in Table 4 shows that LCS performs better than other strategies (i.e. 6 out of 14 cases), while the worst results have been obtained by TVG followed by AETG and IPOG.

Table 5 show that LCS obtains the smallest test suite in almost all cases, only HS has managed to outperform LCS in one case (i.e. case No. 13). Both of LCS and HS produce equal results in 4 cases as marked with bold font. Comparing CS with LCS only, the results in Table 4 show that the results of LCS are much better than standard CS.

Table 4. Comparison with existing strategies ($t \leq 3$)

No.	Systems	t	AETG	IPOG	Jenny	TVG	SA	ACO	GA	LCS
1	3^4	2	**9**	**9**	10	11	**9**	**9**	**9**	**9**
2	3^{13}	2	**15**	20	20	19	16	17	17	18
3	10^{10}	2	NA	176	**157**	208	NA	159	157	**157**
4	5^{10}	2	NA	50	45	51	NA	NA	NA	**42**
5	8^{10}	2	NA	117	104	124	NA	NA	NA	**102**
6	15^{10}	2	NA	373	**336**	473	NA	NA	NA	352
7	3^{10}	2	NA	20	19	18	NA	NA	NA	**16**
8	3^6	3	47	53	51	49	**33**	**33**	**33**	39
9	4^6	3	105	**64**	112	123	**64**	**64**	**64**	101
10	5^6	3	NA	216	215	234	152	**125**	**125**	196
11	6^6	3	343	382	373	407	**300**	330	331	335
12	7^{10}	2	NA	90	83	98	NA	NA	NA	**79**
13	$7^1\ 6^1\ 5^1\ 4^6\ 3^8\ 2^3$	3	45	43	50	51	**42**	**42**	**42**	50
14	$5^2\ 4^2\ 3^2$	3	NA	111	131	136	**100**	106	108	118

Table 5. Comparison with existing strategies

No.	Systems	t	PSO	HS	CS	LCS
1	2^{10}	2	8	**7**	8	8
2	3^{10}	2	17	NA	17	**16**
3	2^{10}	3	17	**16**	**16**	**16**
4	6^6	3	42	**39**	43	**39**
5	5^7	4	1209	1186	1200	**1177**
6	5^8	4	1417	1358	1415	**1329**
7	2^{10}	5	82	81	79	**75**
8	3^7	5	441	NA	439	**437**
9	2^{10}	6	158	158	157	**154**
10	3^7	6	977	NA	973	971
11	2^{10}	7	NS	298	NS	**292**
12	2^{10}	8	NS	498	NS	**501**
13	2^{10}	9	NS	**512**	NS	587
14	2^{10}	10	NS	**1024**	NS	**1024**

Figure 5 shows the comparison results between CS and LCS. LCS strategy appears to generate the optimum results in most cases owing to enhancement of its elitism (i.e. based on learning from other solution instead of purely on random basis).

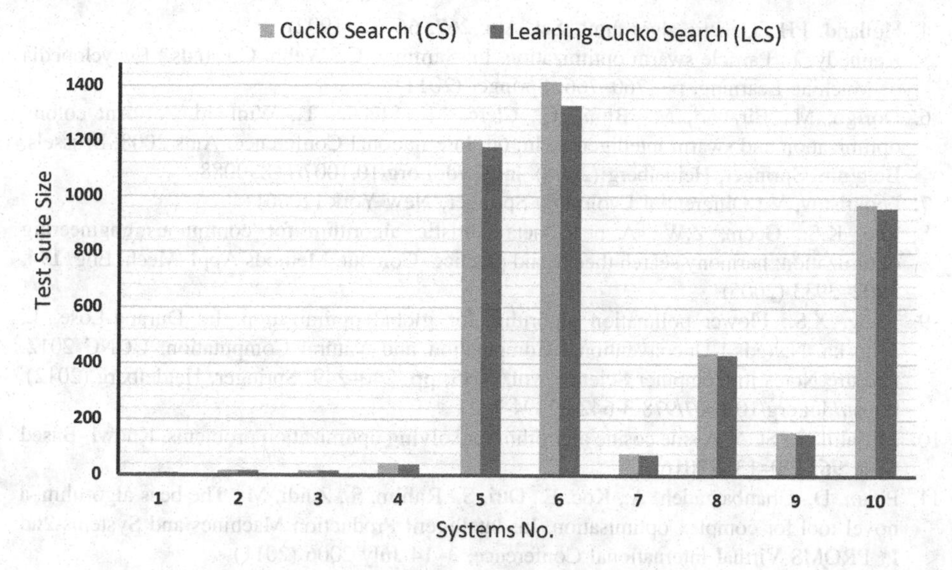

Fig. 5. CS and LCS test suite size comparison

6 Conclusion and Future Work

In this paper, LCS is adopted for design and implementation a new *t*-way strategy for test generation. The strategy utilizes the TLBO's learner phase as part of the Cuckoo elitism process. Initial results show that LCS is able to outperform some of existing meta-heuristic based strategies including origin CS.

Owing to its promising results, we intend to extend this work to apply LCS on different *t*-way interaction possibilities such as cumulative strength interaction, variable strength interaction, and input output based relation. As part of the future work, we plan to improve the design of the LCS by hybridize LCS with another meta-heuristic to improve its overall search capabilities.

Acknowledgment. The work reported in this paper is funded by MOSTI eScience fund for the project titled: Constraints T-Way Testing Strategy with Modified Condition/Decision Coverage from the Ministry of Science, Technology, and Innovation, Malaysia. We thank MOSTI for the contribution and support. Mr. Abdullah B. Nasser is the recipient of the Graduate Research Scheme from Universiti Malaysia Pahang.

References

1. Yang, X.S., Deb, S., Fong, S.: Metaheuristic algorithms: optimal balance of intensification and diversification. Appl. Math. Inf. Sci. **8**, 977–983 (2013)
2. Glover, F., Laguna, M.: Tabu Search. Springer, New York (1999)
3. Kirkpatrick, S.: Optimization by simulated annealing: quantitative studies. J. Stat. Phys. **34**, 975–986 (1984)

4. Holland, J.H.: Genetic algorithms. Sci. Am. **267**, 66–72 (1992)
5. Kennedy, J.: Particle swarm optimization. In: Sammut, C., Webb, G.I. (eds.) Encyclopedia of Machine Learning, pp. 760–766. Springer (2011)
6. Dorigo, M., Birattari, M., Blum, C., Clerc, M., Stützle, T., Winfield, A.: Ant colony optimization and swarm intelligence. In: 6th International Conference, Ants 2008, Brussels, Belgium. Springer, Heidelberg (2008). https://doi.org/10.1007/11839088
7. Feoktistov, V.: Differential Evolution. Springer, New York (2006)
8. Lee, K.S., Geem, Z.W.: A new meta-heuristic algorithm for continuous engineering optimization: harmony search theory and practice. Comput. Methods Appl. Mech. Eng. **194**, 3902–3933 (2005)
9. Yang, X.S.: Flower pollination algorithm for global optimization. In: Durand-Lose, J., Jonoska, N. (eds.) Unconventional Computation and Natural Computation. UCNC 2012. Lecture Notes in Computer Science, vol. 7445, pp. 240–249. Springer, Heidelberg (2012). https://doi.org/10.1007/978-3-642-32894-7_27
10. Mirjalili, S.: SCA: A sine cosine algorithm for solving optimization problems. Knowl.-Based Syst. **96**, 120–133 (2016)
11. Pham, D., Ghanbarzadeh, A., Koc, E., Otri, S., Rahim, S., Zaidi, M.: The bees algorithm–a novel tool for complex optimisation. In: Intelligent Production Machines and Systems-2nd I* PROMS Virtual International Conference, 3–14 July 2006 (2011)
12. Yang, X.-S., Deb, S.: Cuckoo search via lévy flights. In: 2009 World Congress on Nature and Biologically Inspired Computing, NaBIC 2009, pp. 210–214. IEEE (2009)
13. Yang, X.S.: Firefly algorithm, lévy flights and global optimization. In: Bramer, M., Ellis, R., Petridis, M. (eds.) Research and Development in Intelligent Systems XXVI, pp. 209–218. Springer, London (2010). https://doi.org/10.1007/978-1-84882-983-1_15
14. Blum, C., Roli, A.: Hybrid metaheuristics: an introduction. In: Blum, C., Aguilera, M.J.B., Roli, A., Sampels, M. (eds.) Hybrid Metaheuristics. Studies in Computational Intelligence, vol. 114, pp. 1–30. Springer, Heidelberg (2008). https://doi.org/10.1007/978-3-540-78295-7_1
15. He, Z., He, H., Liu, R., Wang, N.: Variable neighbourhood search and tabu search for a discrete time/cost trade-off problem to minimize the maximal cash flow gap. Comput. Oper. Res. **78**, 564–577 (2017)
16. Yıldız, A.R.: A novel hybrid immune algorithm for global optimization in design and manufacturing. Robot. Comput.-Integr. Manufact. **25**, 261–270 (2009)
17. Wang, Z., Rahman, M., Wong, Y., Sun, J.: Optimization of multi-pass milling using parallel genetic algorithm and parallel genetic simulated annealing. Int. J. Mach. Tools Manufact. **45**, 1726–1734 (2005)
18. Alsewari, A.R.A., Zamli, K.Z.: Design and implementation of a harmony-search-based variable-strength t-way testing strategy with constraints support. Inf. Softw. Technol. **54**, 553–568 (2012)
19. Nurmela, K.J.: Upper bounds for covering arrays by tabu search. Discrete Appl. Math. **138**, 143–152 (2004)
20. Stardom, J.: Metaheuristics and the Search for Covering and Packing Arrays. Simon Fraser University, Burnaby (2001)
21. Ahmed, B.S., Zamli, K.Z., Lim, C.P.: Constructing a t-way interaction test suite using the particle swarm optimization approach. Int. J. Innovative Computing, Inf. Control **8**, 431–452 (2012)
22. Chen, X., Gu, Q., Qi, J., Chen, D.: Applying particle swarm optimization to pairwise testing. In: IEEE 34th Annual on Computer Software and Applications Conference (COMPSAC), pp. 107–116. IEEE (2010)
23. Williams, A.: TConfig download page (2008)

24. Lei, Y., Tai, K.-C.: In-parameter-order: a test generation strategy for pairwise testing. In: Proceedings of the 3rd IEEE International Symposium on High Assurance Systems Engineering, pp. 254–261. IEEE Computer Society (1998)
25. Lei, Y., Kacker, R., Kuhn, D.R., Okun, V., Lawrence, J.: IPOG/IPOG-D: efficient test generation for multi-way combinatorial testing. Softw. Test. Verification Reliab. **18**, 125–148 (2008)
26. Lei, Y., Kacker, R., Kuhn, D.R., Okun, V., Lawrence, J.: IPOG: a general strategy for t-way software testing. In: 14th Annual IEEE International Conference and Workshops on the Engineering of Computer-Based Systems (ECBS 2007), pp. 549–556. IEEE (2007)
27. Forbes, M., Lawrence, J., Lei, Y., Kacker, R.N., Kuhn, D.R.: Refining the in-parameter-order strategy for constructing covering arrays. J. Res. Nat. Inst. Stand. Technol. **113**, 287 (2008)
28. Cohen, D.M., Dalal, S.R., Fredman, M.L., Patton, G.C.: The AETG system: an approach to testing based on combinatorial design. IEEE Trans. Softw. Eng. **23**, 437–444 (1997)
29. Lehmann, E., Wegener, J.: Test case design by means of the CTE XL. In: Proceedings of the 8th European International Conference on Software Testing, Analysis and Review (EuroSTAR 2000), Kopenhagen, Denmark (2000)
30. Czerwonka, J.: Pairwise testing in the real world: practical extensions to test-case scenarios. In: Proceedings of 24th Pacific Northwest Software Quality Conference, pp. 419–430. Citeseer (2006)
31. Colbourn, C.J., Cohen, M.B., Turban, R.: A deterministic density algorithm for pairwise interaction coverage. In: The International Association of Science and Technology for Development (IASTED) Conference on Software Engineering, pp. 345–352 (2004)
32. Bryce, R.C., Colbourn, C.J.: The density algorithm for pairwise interaction testing. Softw. Test. Verification Reliab. **17**, 159–182 (2007)
33. https://sourceforge.net/projects/tvg/
34. Zamli, K.Z., Klaib, M.F., Younis, M.I., Isa, N.A.M., Abdullah, R.: Design and implementation of a t-way test data generation strategy with automated execution tool support. Inf. Sci. **181**, 1741–1758 (2011)
35. http://www.burtleburtle.net/bob/math
36. Hartman, A., Klinger, T., Raskin, L.: IBM intelligent test case handler. Discrete Math. **284**, 149–156 (2010)
37. Nie, C., Leung, H.: A survey of combinatorial testing. ACM Comput. Surv. (CSUR) **43**, 11 (2011)
38. Colbourn, C.J., Cohen, M.B., Turban, R.: A deterministic density algorithm for pairwise interaction coverage. In: IASTED Conference on Software Engineering, pp. 345–352. Citeseer (2004)
39. Shiba, T., Tsuchiya, T., Kikuno, T.: Using artificial life techniques to generate test cases for combinatorial testing. In: The 28th Annual International Computer Software and Applications Conference, pp. 72–77 (2004)
40. Ahmed, B.S., Zamli, K.Z.: A variable strength interaction test suites generation strategy using particle swarm optimization. J. Syst. Softw. **84**, 2171–2185 (2011)
41. Ahmed, B.S., Abdulsamad, T.S., Potrus, M.Y.: Achievement of minimized combinatorial test suite for configuration-aware software functional testing using the cuckoo search algorithm. Inf. Softw. Technol. **66**, 13–29 (2015)
42. Cohen, M.B.: Designing Test Suites for Software Interaction Testing. University of Auckland (2004)
43. Shiba, T., Tsuchiya, T., Kikuno, T.: Using artificial life techniques to generate test cases for combinatorial testing. In: International Computer Software and Applications Conference, pp. 72–77. IEEE (2004)

44. Chen, X., Gu, Q., Qi, J., Chen, D.: Applying particle swarm optimization to pairwise testing. In: IEEE 34th Annual Computer Software and Applications Conference, pp. 107–116. IEEE (2010)
45. Nasser, A.B., Alsariera, Y.A., Alsewari, A.R.A., Zamli, K.Z.: Assessing optimization based strategies for t-way test suite generation: the case for flower-based strategy. In: 5th IEEE International Conference on Control Systems, Computing and Engineering, Pinang, Malaysia (2015)
46. Nasser, A.B., Hujainah, F., Alsewari, A.A., Zamli, K.Z.: Sequence and sequence-less t-way test suite generation strategy based on flower pollination algorithm. In: 2015 IEEE Student Conference on Research and Development (SCOReD), pp. 676–680. IEEE (2015)
47. Alsariera, Y.A., Nasser, A.B., Zamli, K.Z.: Benchmarking of Bat-inspired Interaction Testing Strategy
48. Alsariera, Y.A., Zamli, K.Z.: A bat-inspired strategy for t-way interaction testing. Adv. Sci. Lett. **21**, 2281–2284 (2015)
49. Zamli, K.Z., Alkazemi, B.Y., Kendall, G.: A tabu search hyper-heuristic strategy for t-way test suite generation. Appl. Soft Comput. **44**, 57–74 (2016)
50. Zamli, K.Z., Din, F., Kendall, G., Ahmed, B.S.: An experimental study of hyper-heuristic selection and acceptance mechanism for combinatorial t-way test suite generation. Inf. Sci. **399**, 121–153 (2017)
51. Zamli, K.Z., Din, F., Baharom, S., Ahmed, B.S.: Fuzzy adaptive teaching learning-based optimization strategy for the problem of generating mixed strength t-way test suites. Eng. Appl. Artif. Intell. **59**, 35–50 (2017)
52. Rao, R.V., Savsani, V.J., Vakharia, D.: Teaching–learning-based optimization: a novel method for constrained mechanical design optimization problems. Comput. Aided Des. **43**, 303–315 (2011)
53. Liu, X., Fu, M.: Cuckoo search algorithm based on frog leaping local search and chaos theory. Appl. Math. Comput. **266**, 1083–1092 (2015)
54. Zhou, Y., Zheng, H.: A novel complex valued cuckoo search algorithm. Sci. World J. **2013** (2013)
55. Li, X., Yin, M.: A hybrid cuckoo search via Lévy flights for the permutation flow shop scheduling problem. Int. J. Prod. Res. **51**, 4732–4754 (2013)

Harnessing Supremacy of Big Data in Retail Sector via Hadoop

Amarjeet Singh Cheema[(✉)]

Computer Science, Punjabi University, Patiala, (PB), India
amarjeet.chima@gmail.com

Abstract. In today's world of IT revolution, the data generated in most organizations is voluminous, heterogeneous and at high velocity which is popularly known as Big data. Data in the retail industry is increasing exponentially. In today's hyper-competitive sales environment, buyers compare prices on the web and share their experiences on the Internet-good, bad, and neutral. So, retailers are increasingly turning to predictive analytics to fulfill the needs of their customers to get maximum profit. The Large volume of data is generated across their supply chains at point-of-sale and at same time data explosion can be experienced from social media and weblogs. Analysis of this huge heterogeneous data can provide the greater opportunities for retailers to win in today's competitive market. Because of Volume, Velocity, and Variability of this data, it is difficult to handle it by using traditional database management tools. Special tools are used to handle Big Data. Apache Hadoop is one such framework which is capable of handling huge databases via its several components.

This research paper focus on overcoming the hurdles of big data i.e. huge heterogeneous data in the retail sector. In this paper, how to handle structured and unstructured big data is discussed. The main objective of this research paper is to use big data analytic for analyzing retail data to better understand customers in a systematic manner, so that retailer can take better decisions.

Keywords: Big data · Customers · Retailer · Hadoop · Retail industry

1 Introduction

In today's world of IT revolution, the data generated in most organizations is voluminous, heterogeneous and at a very high velocity which is popularly known as Big data. Big data refers to the data sets that are too big to handle by using the conventional database management tools and are emerging in many important applications, such as retail industry, business informatics, social networks, social media, genomics, and meteorology [5]. Big data itself is of no use as it is very large and complex one has to refine it to get a meaningful information. Analyzing big data helps analysts, business users and researchers to make better and faster decisions, using data which was not accessible previously.

Retail Industry plays a vital role in any economy. Retail data together with retail analytics can assist retailers in understanding the customer expectations and responding them with appropriate action, by finding new and faster ways to identify product

R. Sharma et al. (Eds.): ICAN 2017, CCIS 805, pp. 111–123, 2018.
https://doi.org/10.1007/978-981-13-0755-3_9

preferred by consumers. The purchasing decision journey for consumers involves multiple steps, which are now being captured, digitized, and transformed into metrics and data. Now focus is on how to extract information from data that can be turned into a competitive advantage for the retailer and a better shopper experience for the consumer.

This research paper focus on using the big data analytic for analyzing retail data to better understand customers in a systematic manner. We also discussed how to handle structured and unstructured big data by using Hadoop. This research paper is organized as follows: Big data and why it is important is discussed in Sect. 2. In Sect. 3 Hadoop and its components are illustrated. Data and analysis methodology is discussed in Sect. 4. In Sect. 5 experimental setup and overview of tools used, are illustrated. Data analysis and results are presented in Sect. 6 and finally, conclusions are summed up in Sect. 7.

2 Big Data

Big data is the term used for massively large data set which includes different variety of data which may be both structured or unstructured and arriving at very high speed. Big data can't be handled by common data management tools [5]. Data is called big data of its characteristics of Volume, Variety, Velocity, Variability, Veracity. Big data is important because of its characteristics i.e. more data leads to more accurate analyses, more accurate analyses lead to better decision making, better decision means greater operational efficiencies, cost reductions and reduced risk (Fig. 1).

Few years ago, in the past technology platforms were built only to handle either structured data or unstructured data. But now days every organization have heterogeneous data i.e. structured and unstructured data. So they need a powerful computing

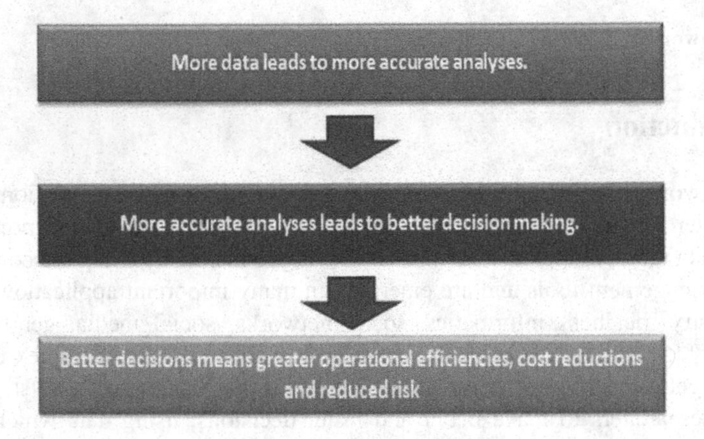

Fig. 1. Importance of big data

environment to process this large disparate data. Organizations have large repositories of unstructured data in the form of documents, emails etc. and also structured information corporate systems like Oracle, SAP, NetSuite etc. Today's organizations, however, are utilizing, sharing and storing more information in varying formats, including:

- e-mail and Instant Messaging
- Collaborative Intranets and Extranets
- Public websites, wikis, and blogs
- Social media channels
- Video and audio files
- Data from industrial sensors, wearables and other monitoring devices

These formats include disparate data. Unstructured data in organizations adds up to as much as 85% of the information that businesses store. The ability to extract high value of information from this data to enable innovation and competitive gain is known as Big Data analytics. The main objective of big data analytics is to help organizations to make best decisions by enabling data scientists, business users, researchers and other analytics professionals to analyses large volumes of data that may be inaccessible by conventional database tools. The business users working with big data basically want to extract a useful information from big data. This extracted information can lead to better customer service, new revenue opportunities, operational efficiency, effective marketing, competitive advantages over rival organizations and other business benefits.

2.1 Applications of Big Data Analytics in Various Fields

Big data analytics become important because more data you have more information you can extract. Now day's data around the world is growing exponentially creating both opportunity and challenges for the data analysts. It is estimated that data produce in 2020 will be 44 times greater than it was in 2009. [34] According to experts, the digital information by 2020 will grow to 40 zettabytes, which is now 3.2 zettabytes. The data that could be generated from beginning of life to 2013 can now be generated only in two days [35].

Big data analytics is used in various field throughout the world for getting best results. In financial institutions big data analysis can be used for identifying fraud and preventing from further damage. Governments used big data analytics to increase their security and for prevention from outside cyber threats. Big data analytics is used in healthcare industry to cure disease and improve quality of life. Telecommunications companies utilize big data analytics to recommend the offers to the customers and also plan the best ways to manage networks. Special tools are used for these tasks of big data analytics. Apache Hadoop is one such framework which is capable of handling huge databases via its several components.

3 Hadoop

Big Data, because of its volume, veracity, variety and velocity can't be handled by traditional database tool. So, a special tool is needed to handle Big Data known as Apache Hadoop. Doug Cutting and Mike Cafarella took the solution provided by Google and started an Open Source Project called HADOOP in 2005 Cutting named it after elephant which was his son's toy. Hadoop is a registered trademark of the Apache Software Foundation. The Hadoop framework is used by IBM, Yahoo and Google, largely for applications involving search engines and advertising [14].

Hadoop is open source framework consists of subprojects around distributed computing. The various functions are performed by various subprojects. They can be utilized according to needs. Hadoop mainly consists of Core, Hadoop distributed file system, MapReduce. Other subprojects can be added according to need.

Core: The main motive of Hadoop core is to provide the access to distributed file system and general input-output (Serialization, Java RPC, Persistent data structures). Core is made up of interface components which help the user to interact with internal components of Hadoop. Core component also takes account of minimizing network traffic between the servers which is called Rack Awareness.

Hadoop Distributed File System: HDFS is a storage system of Hadoop. Hadoop Distributed File System as its name suggests it is a distributed file system that distributes the data to its nodes for fast computation by principle of multi-computing. It also made multiple replicas of data blocks to which helps to minimize the risk of system failure [9].

MapReduce: Hadoop MapReduce is a programming model which works on divide and conquer concept. It is used for algorithms that process huge amounts of data on distributed clusters of computer nodes simultaneously. For accessing files and to store reduce result HDFS is used [12] (Fig. 2).

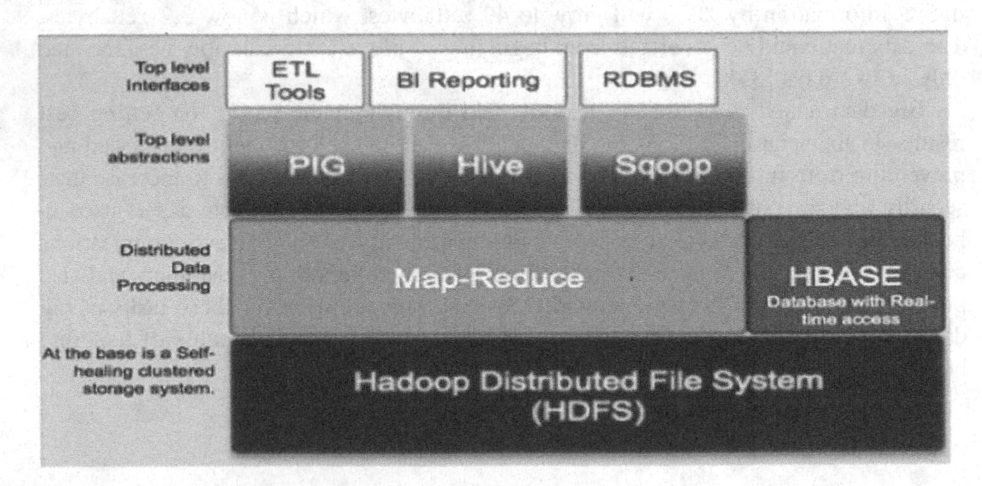

Fig. 2. Hadoop architecture

The other sub-projects of Hadoop are:

HBase: HBase is column oriented distributed data base build on Hadoop distributed file system. HDFS data is mapped by HBase providing Java API access to Database. Main motive of HBase is to host the large tables running on top of Hadoop [20].

Pig: Pig is a scripting language. The main feature of pig is that their structure can be parallelized which intend helps to handle huge data sets, and their built-in functionality leads to easy programming as compared to traditional Java MapReduce Jobs [19].

Zookeeper: Zookeeper gives the services of maintaining the configuration naming and information. It provides group service and also helps in distributed synchronization [18].

Hive: Hive is query language which helps in refining the large data sets. The use familiar with SQL can easily query Hive [16].

Chukwa: Chukwa is used for monitoring clusters of systems. Chukwa has a powerful toolkit for analyzing results to make the best use of data [25].

HCatalog: Table representation of data is shown in HCatalog. Data is represented in relational view for its abstraction [24].

4 Data and Analysis Methodology

As retail industry had entered the era of big data. Which means the size of data sets are too large (Volume), heterogeneous including structure, semi-structure, and unstructured data (Variety), arriving at a very high rate (Velocity) than before in the past. To refine this type of data, one needs a smart and powerful data analytic.

This section presents the proposed approach for retail big data analysis. In this Hadoop is used to handle Big data. Retails Big Data for research consists of both structured and unstructured data, along with point of sale (POS) data. The proposed approach works in two phases. In the first step, Big data is processed with Hadoop and in the second step, the processed data is shown in visuals by using business intelligence tool. This approach can be demonstrated as given below.

- Raw Big data is processed by using Hortonworks platform Hadoop (Hortonworks HDP sandbox). Large (volume) heterogeneous (variety) data sets are refined by implementing the algorithms in Hive.
- After dealing with Big Data, the data is exported to BI tool (Microsoft excel 2016) for further analysis.
- Microsoft excel 2016 is used for further analyzing the data and to show the visualizations.

The whole process of analyzing the retail big data from processing through Hadoop to visualization in business intelligence tool is shown in the flow diagram in Fig. 3.

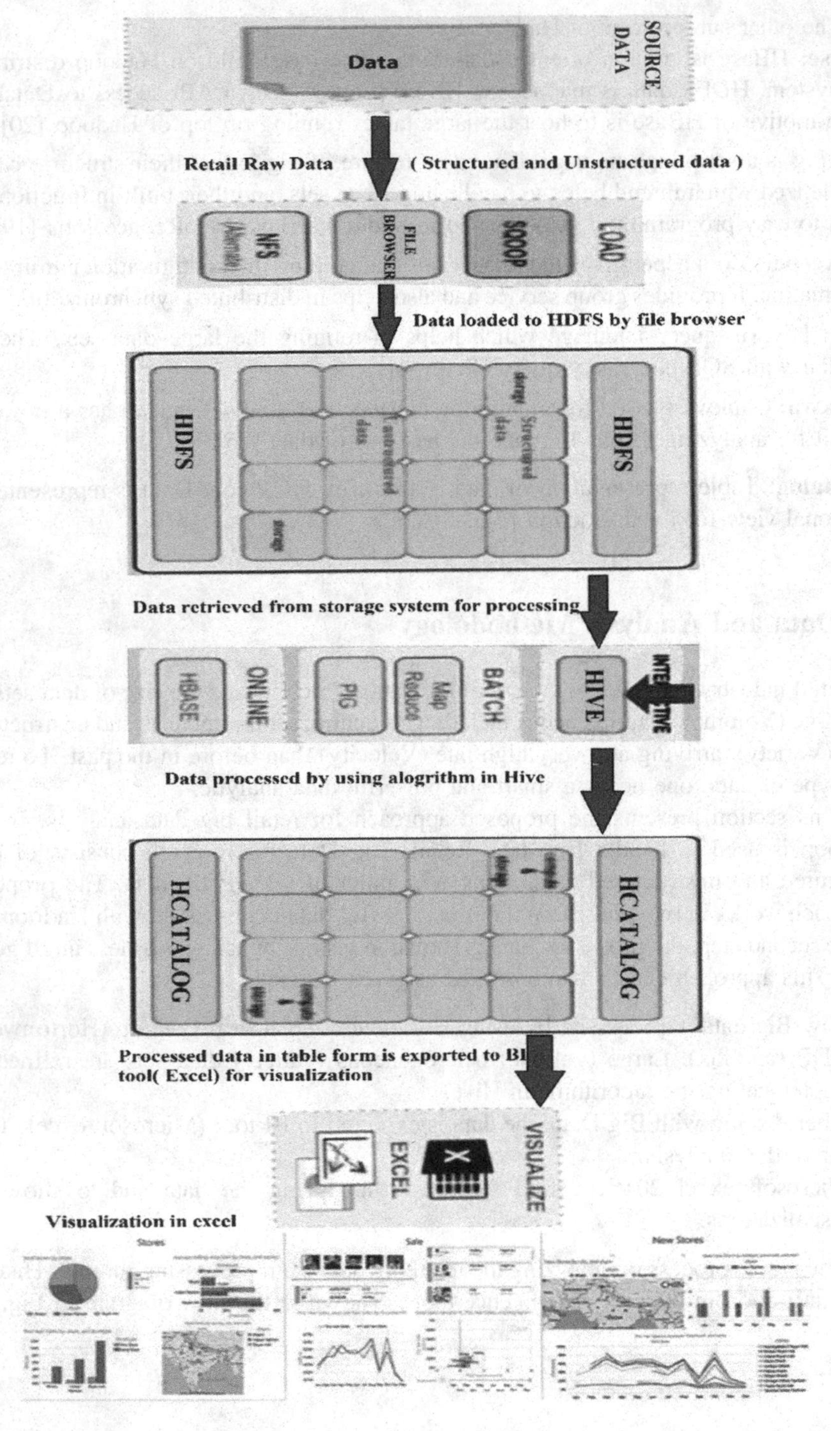

Fig. 3. Flow diagram of experiment

5 Experimental Setup

This section will introduce the results of retail big data analysis. Our results are all calculated on a desktop computer with an Intel Core i5 3.4 GHz CPU, 6 GB RAM and NVIDIA Quadro 2000 GPU. We have used Hortonworks Sandbox on Window 10. A 64-bit machine which supports virtualization is needed to run Hortonworks Sandbox. We have used VMware for the sandbox. This experiment was performed on large datasets which mainly contains both structured and unstructured texts data.

5.1 Overview of Tools and Technologies Used

Hortonworks HDP Sandbox
Hortonworks HDP sandbox is a bundled set of applications, configuration settings, and operating system that work together as a signal unit. It a is virtual appliance that runs with virtual machine. For running Hortonworks sandbox, we must have virtual machine environment on our host machine. We can install Oracle Virtual Box or VMware Player. After processing the data, it can be sent to any Business Intelligence tool for further analysis by using ODBC Driver [31].

VMware Player
VMware Player is a free desktop application that helps us to run a virtual machine on Windows or Linux. VMware Player provides user interface for running preconfigured virtual machine created with VMware Workstation, GSX server and ESX server. VMware Player includes features that let us configure the virtual machine for optimal performance and take advantage of host machine device. In this research VMware is used to run Hortonworks sandbox.

Hive ODBC Driver
Hive ODBC Driver enables us to export the data from Hadoop to any BI tool, which are used to analyses and visualize the data refined by Hadoop.

Microsoft Excel 2016
Microsoft excel is a powerful Business Intelligence tool. Excel versions above excel 2013 support Power pivot to manage data model and Power view for visualization. In this research, Microsoft excel 2016 is used for visualization of data imported from Hadoop.

6 Data Analysis and Results

This section relates to the data analysis of retail Big Data, which was collected from different sources from the internet. The interpretations are mentioned so that meaningful recommendations and conclusion can be drawn. The analysis was performed by using Big Data analytic tool Hadoop (Hortonworks HDP Sandbox) and Business Intelligence tool (Microsoft Excel). Hadoop is used for handling Big Data and creating a relational database. The business Intelligence tool is used to further analyses the retail relational database and showing the visualizations for maximizing the benefits of retailers. The samples of raw data to be processed are shown in Figs. 4 and 5.

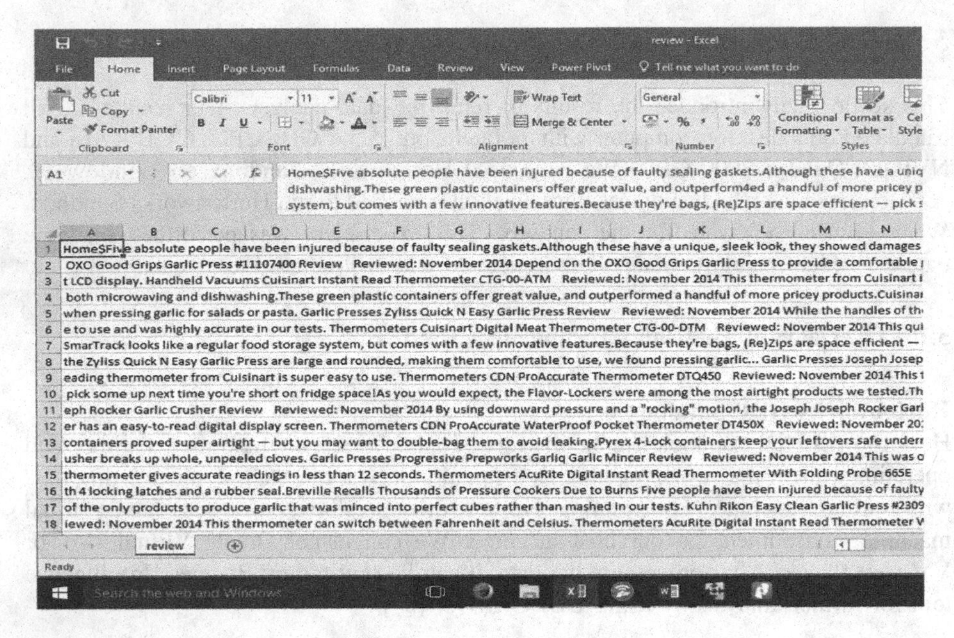

Fig. 4. Unstructured data

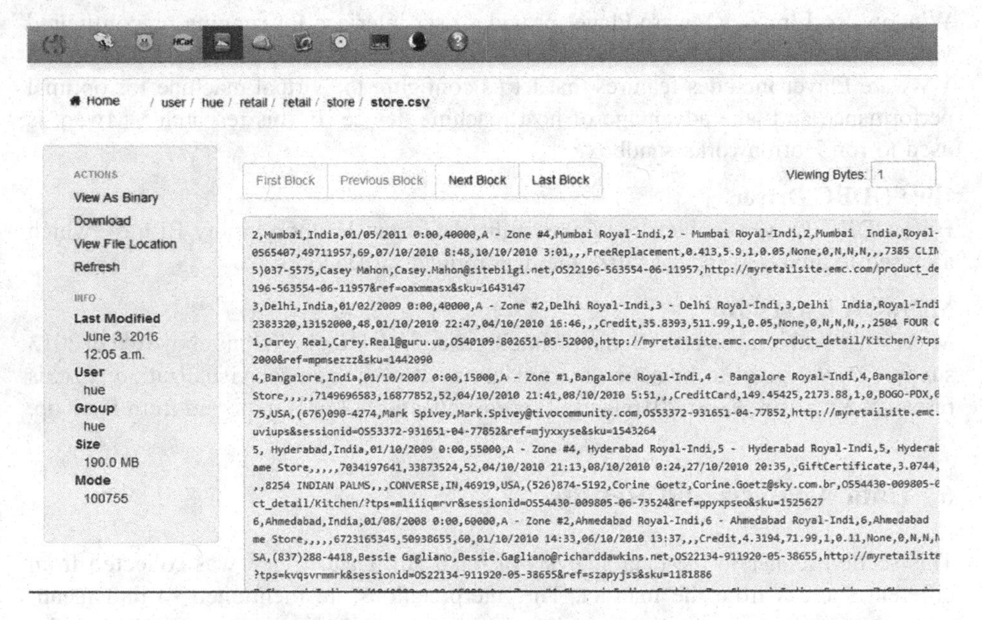

Fig. 5. View of unrefined data in Hadoop

We had processed the retail big data by using Big Data Analytic tool Hadoop i.e. Hortonworks HDP sandbox. For processing the large structured and unstructured data, algorithms are implemented in Hive. By using Hive, the large and heterogeneous data of retails is converted into relational data tables. Some examples of Apache Hive scripts for different purposes are illustrated as under.

- To create a table "retail" from database "/user/retail"

```
Create External Table retail
(
  Month_ID                        bigint
, Item_ID                         bigint
, Location_ID                     bigint
, Sum_GrossMargin                 double
, Sum_Regular_Sales_Rupees        double
, Sum_Markdown_Sales_Rupees       int
, ScenarioID                      int
, ReportingPeriodID               bigint
, Sum_Regular_Sales_Units         int
, Sum_Markdown_Sales_Units        int
)
Row format delimited fields Terminated by ',' stored as
Textfile Location '/user/retail';
```

- To count number of different manufacturers in database "retail"

```
Select count (distinct manufacturer) from retail;
```

- To count any number of items of particular category

```
Select word, COUNT (*) from retail Lateral view explode
(split (item,' ')) Ltable as word Group by word;
```

Similarly, user can write the number of different scripts and queries and obtain useful results. Figure 6 shows the hive script written in beeswax of Hortonworks sandbox for processing of data.

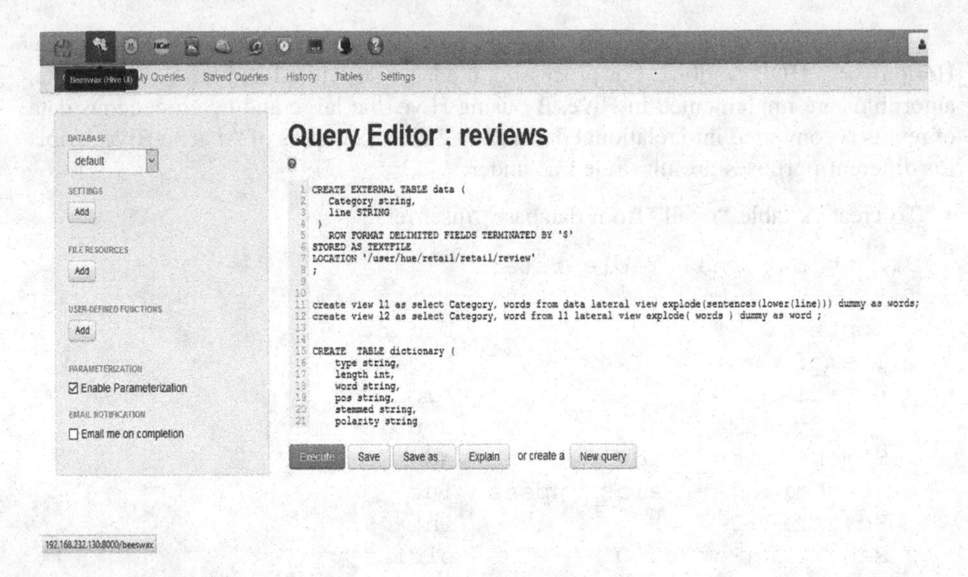

Fig. 6. Hive script

Here the unstructured data is refined and converted into a relational table. Data in Fig. 4 is processed and converted into a relational table as shown in Fig. 7. Also, data of Fig. 5 is processed and results are shown in Fig. 8.

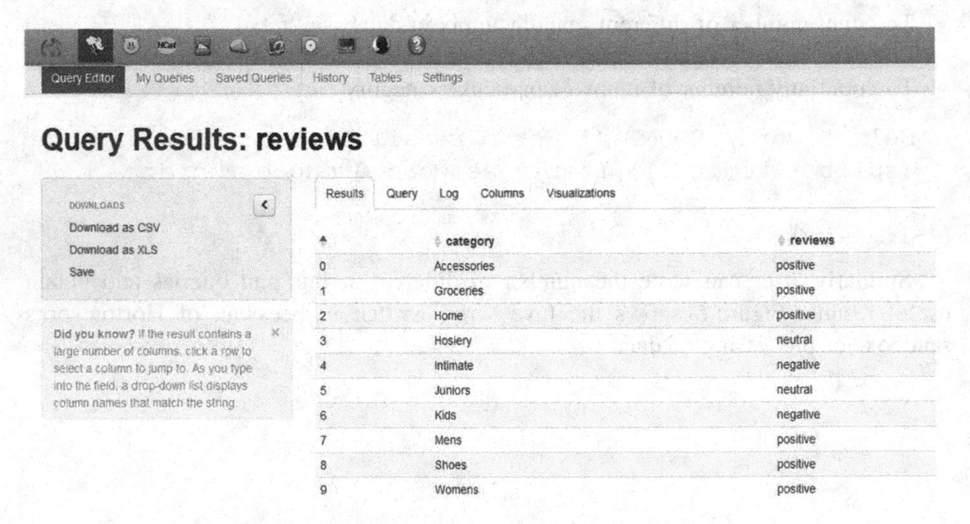

	category	reviews
0	Accessories	positive
1	Groceries	positive
2	Home	positive
3	Hosiery	neutral
4	Intimate	negative
5	Juniors	neutral
6	Kids	negative
7	Mens	positive
8	Shoes	positive
9	Womens	positive

Fig. 7. Unstructured data converted to structured data

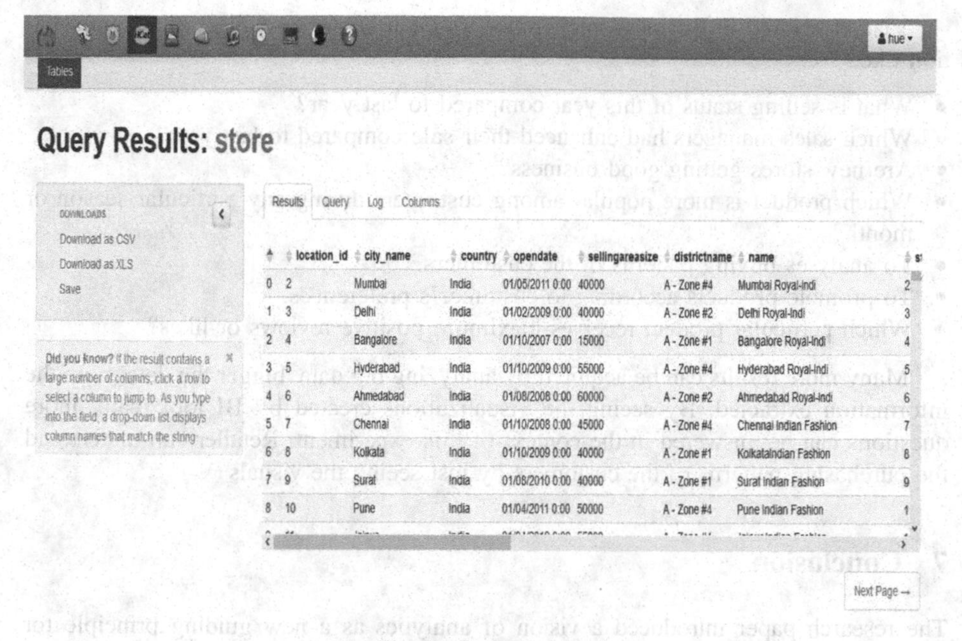

	location_id	city_name	country	opendate	sellingareasize	districtname	name
0	2	Mumbai	India	01/05/2011 0:00	40000	A - Zone #4	Mumbai Royal-Indi
1	3	Delhi	India	01/02/2009 0:00	40000	A - Zone #2	Delhi Royal-Indi
2	4	Bangalore	India	01/10/2007 0:00	15000	A - Zone #1	Bangalore Royal-Indi
3	5	Hyderabad	India	01/10/2009 0:00	55000	A - Zone #4	Hyderabad Royal-Indi
4	6	Ahmedabad	India	01/08/2008 0:00	60000	A - Zone #2	Ahmedabad Royal-Indi
5	7	Chennai	India	01/10/2008 0:00	45000	A - Zone #4	Chennai Indian Fashion
6	8	Kolkata	India	01/10/2009 0:00	40000	A - Zone #1	KolkataIndian Fashion
7	9	Surat	India	01/08/2010 0:00	40000	A - Zone #1	Surat Indian Fashion
8	10	Pune	India	01/04/2011 0:00	50000	A - Zone #4	Pune Indian Fashion

Fig. 8. Refined data

After processing the Big data with Hortonworks Hadoop, it can be exported to any Business Intelligence tool for further analysis by using ODBC Driver. The processed data is exported to Microsoft excel 2016 for further analysis and visualization. In Microsoft excel visualizations are created in power view. Figure 9 shows the visualizations created in our research by BI tool Microsoft excel 2016.

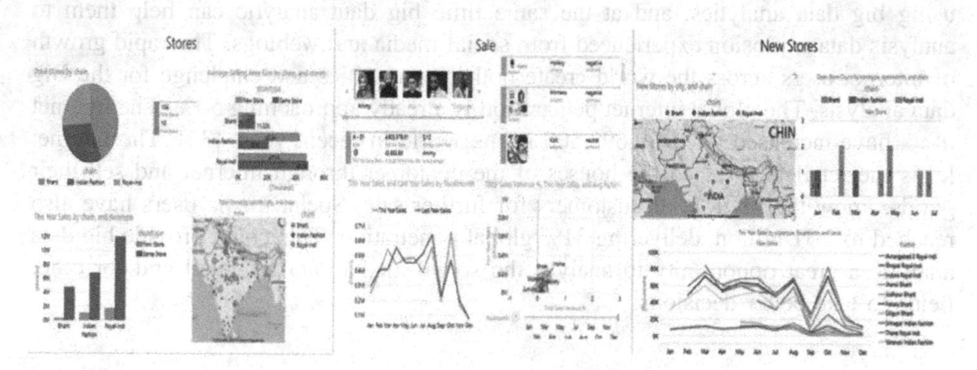

Fig. 9. Visuals created in microsoft excel

Visuals created in Business intelligence tool can help the retailers to get information like:

- What is selling status of this year compared to last year?
- Which sales managers had enhanced their sale compared to last year?
- Are new stores getting good business?
- Which product is more popular among customers during any particular season or month?
- To analyses buying patterns of the customers.
- To promote products according to customer's preferences.
- Which particular product receives maximum positive reviews or likes?

Many more results can be acquired by analyzing big data, bigger the data-more the information extracted. By seeing the visualizations created by BI tool many more questions can be answered in the context of this experiment. Retailer can understand the purchasing patterns of the customers by just seeing the visuals

7 Conclusion

The research paper introduced a vision of analytics as a new guiding principle for operating in today's tumultuous retail environment and discussed the power of becoming a data driven decision-making culture, and shown how to access accurate, scalable, and actionable data that can help retailers to set a roadmap to success through a better understanding of their customers and their store operations. The paper also shows that how large heterogeneous data can be handled by using Hadoop. We've also covered how data can reveal exposures as well as opportunities for the retailer. The right insights enable a closer, stronger relationship with consumers. As retailers, can analysis large volume of data generated across their supply chains at point-of-sale by using big data analytics, and at the same time big data analytic can help them to analysis data explosion experienced from social media and weblogs. The rapid growth of internet users across the world create both opportunities and challenge for the Big data analysts. The global internet penetration is already approaching 50%. The internet users have increased to 3,739,698,500 in the world in recent years [32]. The internet leads the retailers to reach the houses of the customer through internet and sell their goods, know the reviews of customers for further sale. Social media users have also reached to 2.31 billion, delivering 31% global penetration. This could provide big data analytic a great opportunity to analyze the social media data for retail and for other fields to take better decisions.

References

1. Manikandan, S.G., Ravi, S.: Big Data analysis using apache Hadoop. IEEE (2014)
2. Nallasamy, S., Marimuthu, R.: Big Data – Relevance for Retailers
3. IBM, Analytics: The real-world use of big data (2013)
4. Wipro, The Data Strom: Retail and big data revolution (2013)

5. Big-data. http://www.sas.com/en_ca/news/sascom
6. Retails. https://en.wikipedia.org/wiki/Retail
7. Big data. https://en.wikipedia.org/wiki/Big_data
8. Srinivasan, N., Nayar, R.: Harnessing the Power of big data Big opportunity for retailers to win customers (2013)
9. Hadoop. http://www.sas.com/en_my/insights/big-data/hadoop.html
10. Big data analytics. http://www.webopedia.com/big_data_analytics.html
11. Chen, H., Chiang, R.H.L., Storcy, C.: Business Intelligence and Analytics: From Big Data to Big Impact (2012)
12. MapReduce. https://en.wikipedia.org/wiki/MapReduce
13. Gutierrez, D.D.: Inside Big Data guide to retail (2015)
14. Hadoop. http://hadoop.apache.org
15. Howe, K.: Beyond Big Data: How Next-Generation Shopper Analytics and the Internet of Everything Transform the Retail Business (2014)
16. Hive. http://www.tutorialspoint.com//hive/hive_introduction.html
17. Intel: getting started with big data Analytics in retail (2014)
18. Zookeeper. http://www.tutoralspoint.com/zookeeper/zookeeper_overview
19. Pig. http://www.tutorialspoint.com/apache_pig/apache_pig_overview.html
20. Hbase. http://www.tutorialspoint.com/hbase/hbase_overview.html
21. Hadoop. http://www.docfoc.com
22. Big data analytics. http://www.qubole.com/resources/articles/big-data-analytics
23. Big data analytics. http://www.linkedin.com/pulse/big-data-analytics-mario-ruggiero
24. HCatalog. http://www.tutorialspoint.com/hcatalog/hcatalog_introduction.html
25. Hadoop. http://www.saphanatutorial.com/hadoop-online-training-hadoop-basics
26. Hadoop. http://www.forbes.com
27. Katal, A., Wazid, M., Goudar, R.H.: Big data: issues, challenges, tools and good practices, pp. 404–409. IEEE (2013)
28. Kaisler, S., Armour, F., Espinosa, J.A., Money, W.: Big data: issues and challenges moving forward. In: International Conference on System Sciences, pp. 995–1004. IEEE Computer Society
29. Smolan, R., Erwitt, J.: The Human Face of Big Data. Sterling Publishing Company Incorporated (2012)
30. Ward, J.S., Barker, A.: Undefined by Data: A Survey of Big Data Definitions. School of Computer Science, University of St. Andrews, UK (2013)
31. Hadoop. http://hortonworks.com
32. Internet users. http://www.internetworldstats.com
33. Big data analytic. http://www.slideshare.net/economistintelligenceunit/the-data-storm
34. How Entrepreneurs Are Winning by Understanding Big Data. https://www.forbes.com
35. Top 10 Amazing Facts to Know About Big Data. http://www.business2community.com

Intelligent Modelling of Moisture Sorption Isotherms in Milk Protein-Rich Extruded Snacks Prepared from Composite Flour

A. K. Sharma(✉), N. R. Panjagari, and A. K. Singh

ICAR-National Dairy Research Institute, Karnal 132001, Haryana, India
a.sharma@icar.gov.in, ndri.pnr@gmail.com,
aksndri@gmail.com

Abstract. In this paper, connectionist models have been investigated empirically to predict adsorption isotherms of milk protein-rich extruded snacks prepared from composite flour, at different temperatures (*i.e.*, 28, 37 and 45 °C) and water activities (*i.e.*, in the range: 0.112–0.971). These models were based upon error back propagation learning algorithm supplemented with Bayesian regularization optimization mechanism as well as with various combinations/settings of network parameters. In all simulation experiments, the connectionist models with single hidden layer were found to fit the best to the adsorption isotherms data. The best configuration of the connectionist models comprised 10 neurons in the hidden layer with tangent-sigmoid transfer function; which attained accuracy in the range of 0.467–0.958 root mean square percent error (%RMS). Also, several conventional mathematical sorption models including two-parameter models, *viz.*, Lewicki-I, Mizrahi and Modified BET; and three- and four-parameter models, *i.e.*, Ferro-Fontan, GAB, Lewicki-II, Modified GAB, Modified Mizrahi and Peleg were developed for the purpose. The Ferro-Fontan and Peleg were the best similar models among the conventional sorption models, with %RMS lying in the ranges: 1.63–1.89 and 1.41–3.33, respectively, for the same temperatures and water activities range. Evidently, the connectionist sorption models developed in this study were found to be superior over conventional sorption models, to efficiently and intelligently predict adsorption isotherms of milk protein-rich extruded snacks prepared from composite flour.

Keywords: Adsorption isotherms · Connectionist models
Empirical sorption models · Extruded snacks · Predictive analytics

1 Introduction

1.1 Connectionist Models – An Intelligent Paradigm

A connectionist model, also known as Artificial Neural Network (ANN) model, is the information processing paradigm that is inspired by the way biological nervous systems, such as the human brain processes information. The main component of this paradigm is novel structure of information processing system. It is comprises several highly interconnected processing elements (also called artificial neurons, nodes or units in

R. Sharma et al. (Eds.): ICAN 2017, CCIS 805, pp. 124–137, 2018.
https://doi.org/10.1007/978-981-13-0755-3_10

subsequent discussions) working in unison to solve specific problems. Connectionist models, like human beings, learn by example! A connectionist model is configured for a specific application, such as pattern recognition or data classification, through a learning process. Learning in biological systems involves adjustments to the synaptic connections that exist between the neurons. This is true of connectionist models as well.

Connectionist models, with their amazing ability to derive meaning from complicated or imprecise data, can be used to extract patterns and detect trends that are too complex to be noticed either by experts or conventional computational/statistical techniques. A trained connectionist model can be thought of as an expert in the category of information it has been given to analyze. This expert system can then be used to provide projections about given new situations of interest and answer 'what if' questions. Other advantages include adaptive learning, self-organization, real-time operation and fault tolerance via redundant information coding. However, some network capabilities may be retained even with major network damage!

1.2 Milk Protein-Rich Extruded Snacks Prepared from Composite Flour

Snack foods (or snacks) are defined as a light meal consumed between regular meals; and include a variety of products that can take many forms such as, potato chips or cereal-based snacks [1]. Snacks have a wide variety of shapes, flavors and colors. The increasing demand for snack food is attributed to several factors such as changing lifestyles; people spending more time at work and less time for food preparation; increased income; growing convenience food markets; *etc*. India too, is not new to consumption of snacks. The Indian snacks market has grown subsequently; and the industry continues to exhibit high growth rate.

However, currently, snack market is predominantly fat- and calorie-rich and deficient on nutritional front; thus, selling obesity, whilst aiming at children in particular. To counter this subject palatably, food scientists of Dairy Technology Division at ICAR-National Dairy Research Institute, Karnal (India) have made efforts by way of implementing extrusion cooking for designing a crisp snack by utilizing associated benefits of sweet potato flour (rich in carotenes), barley flour (rich in fiber), rice flour (rich in starch) as ingredients, and rennet casein as the source of milk protein [2].

1.3 Moisture Sorption Isotherms

The importance of water to life is well recognized as well as plays vital role in controlling the growth of microorganism(s) in various agro-food and other biological systems. The simple way for expressing such a status is known as 'water activity'. Water activity (a_w) is a physical measure of active water available in foods. This active water is responsible for the growth of spoilage bacteria, chemical reactions and enzymatic activities. In case of moisture sensitive dairy products especially dehydrated products, water is most important characteristic that affects the physical, chemical, microbial and sensory properties. Generally, food when kept in different relative humidities absorbs or desorbs moisture depending on their water activity. When water activity is less than surrounding humidities, the product will adsorb moisture; and when reverse is the case, it will desorb the moisture.

Moisture sorption isotherm (MSI) illustrates the steady-state amount of water held by the food solids as a function of a_w at constant temperature. There are many important applications of water sorption isotherms in Food Science and Technology. They are mainly used to control deterioration during food storage. Thus, the MSI is an extremely valuable tool for food scientists and engineers as it can be used to predict, which reactions will decrease stability at given moisture. It allows for ingredient selection to change a_w to increase stability; and can be used to predict moisture gain or loss in a package with known moisture permeability. Sorption isotherm is the plot of equilibrium moisture content of a material subjected to different relative humidities at a given temperature in closed environment. Depending on the direction of water transferred the process is named as adsorption or desorption.

An adsorption isotherm can be obtained by subjecting a relatively dry material to an atmosphere of constant relative humidity and measuring the weight gained at equilibrium due to water and repeating this procedure for different relative humidities. Desorption isotherm is found by placing the wet material into an atmosphere of lower relative humidity and then again determining its Equilibrium Moisture Content (EMC) [3].

1.4 Review of Literature

In recent years, the connectionist models have gained wide acceptance in food engineering for predictive analytics. The following studies have investigated prediction potential of connectionist models to study sorption isotherms of various food materials.

Myhara *et al.* [4] determined water sorption isotherms for two date varieties at 15, 25 and 45 °C. Water sorption isotherm was modeled using Guggenheim-Anderson-de Boer (GAB) equation, a modified GAB equation and connectionist model. The GAB equation had a lower Root Mean Square Percent Error (%RMS), *i.e.*, approximately 7% than the modified GAB equation (approximately 16%) in predicting EMC. The connectionist model, optimized through trial and error, was superior to both the GAB equations.

Myhara and Sablani [5] reported that chemical compositional data played a major role in influencing water sorption behavior of dried fruit. Standard theoretical and empirical modelling methods cannot adjust for compositional varietal differences, because the modeling methods relied upon physical measurements only. When chemical compositional data were combined with physical data, through connectionist model, significant improvements in the prediction of water sorption behavior were achieved.

Panchariya *et al.* [6] developed connectionist model using Error Back Propagation (EBP) algorithm based upon Levenberg-Marquardt (LM) method for modeling desorption isotherms of black tea with temperature varying in the range 25–80 °C and the water activity in range of 0.10–0.90. The connectionist model exhibited higher prediction accuracy than the well known phenomenological GAB model, also developed in the study for comparison. It was reported that connectionist model supplemented the available information on the full sorption curves and provided a better understanding of the interaction of black Darjeeling CTC tea particles with water.

Peng *et al.* [7] determined adsorption and desorption isotherms for corn starch powder at 30, 45 and 60 °C with connectionist model based on EBP algorithm *vis-à-vis* conventional mathematical models. The average %RMS of sorption connectionist model was 2.96 and desorption model was 4.84 whereas that of GAB (best among conventional models) was 6.1 and 6.6 at 45 °C. Hence, the connectionist model not only accommodated temperature and water activity parameter, but also was found to be more accurate than conventional mathematical models.

Araromi *et al.* [8] studied physical properties such as EMC, Equilibrium Relative Humidity (ERH), initial moisture content and critical moisture content in corn flour. These were experimentally determined at four different temperatures: 20–50 °C and at various ERH values ranging between 0.113 and 0.9812. Two models, *i.e.*, neural-fuzzy model and GAB model were used to fit experimental isotherm data. The two models were used to predict shelf life of corn flour. Comparisons were made to determine validity and suitability of the two models. Within the range of temperatures investigated, GAB and Neural-fuzzy models better described the experimental data for corn flour for adsorption isotherm. The %RMS of sorption neural-fuzzy model were 2.35, 2.14, 2.12 and 2.02 for 20, 30, 40 and 50 °C, respectively. The GAB model though considered most versatile sorption model available in literature had %RMS of 5.96, 5.82, 5.5 and 5.28 at 20, 30, 40 and 50 °C, respectively. Neural-fuzzy model not only accommodated temperature and water activity parameter but was also better than GAB model in the prediction of corn flour shelf life.

Kingsly and Ileleji [9] studied MSI of corn dried distillers grains using connectionist models. Janjai *et al.* [10] developed connectionist model trained using EBP algorithm to predict the equilibrium moisture content of longan. Chayjan [11] investigated thermodynamic models and connectionist models for computation of sesame seed dehydration energy. Chayjan and Esna-Ashari [12] explored connectionist models *vis-à-vis* empirical models (*i.e.*, D'Arsy-watt, GAB, Halsey, Henderson, Oswin and Smith) for prediction of EMC in raisin. Correa *et al.* [13] studied desorption isotherms and thermodynamic properties of coffee obtained from different processing stages during its drying process by using connectionist model besides conventional models like GAB and modified Henderson models.

Chayjan and Esna-Ashari [14] developed connectionist models *vis-à-vis* four empirical mathematical models, *i.e.*, GAB, Halsey, Henderson and Oswin for estimation of EMC of the dried grape (black currant). All these studies showed that the EMC of different food material investigated were more accurately predicted by connectionist models than by the empirical models.

Gazor and Eyvani [15] determined MSI of red onion slices at 30, 40, 50, and 60 °C over a range of relative humidity from 0.11 to 0.83. The experimental sorption curves were fitted by seven empirical equations: GAB, modified BET, modified Chung–Pfost, modified Halsey, modified Henderson, modified Oswin and modified Smith. Also, three connectionist models were investigated to predict the EMC of onion slices. The modified Oswin model was found acceptable for predicting adsorption moisture isotherms and fitting to the experimental data with %RMS of 15.02. Besides, connectionist model with four layers (2-17-14-1) was selected as the best model for estimation of onion slices' EMC.

Morse *et al.* [16] explored the possibility to use a single connectionist model to predict a wide array of standard adsorption isotherm behavior. It was found that the connectionist models were highly effective to represent adsorption isotherms at a constant temperature. Results showed that a single connectionist model with a hidden layer having three neurons, including the bias neuron, was able to represent very accurately the adsorption isotherm data.

Sharma *et al.* [17, 18] investigated soft computing[1] approach (connectionist models and Adaptive Neuro-Fuzzy Inference System hybrid models) to model adsorption and desorption isotherms for milk and pearl millet-based weaning food called fortified Nutrimix, at four temperatures, 15, 25, 35, and 45 °C over the water activity range: 0.11–0.97.

Recently, Sharma and Sawhney [19] developed six connectionist models to predict sorption (adsorption and desorption) characteristics at three temperatures, *i.e.*, 25, 35 and 45 °C over a water activity range of 0.11–0.97 in dried acid casein prepared from buffalo skim milk. Also, several conventional empirical sorption models were used for fitting the sorption data. The EBP learning algorithm with Bayesian Regularization (BR) and LM optimization techniques as well as various combinations of connectionist network parameters were employed. The results revealed that the connectionist models outperformed the conventional empirical models and, generally, best described the experimental adsorption data for dried acid casein prepared from buffalo skim milk.

The foregoing review of relevant literature revealed that the soft computing models, generally, outperformed the conventional empirical sorption models, and best described the experimental sorption data for various food materials. However, modeling of MSI in milk protein-rich extruded snacks prepared from composite flour, with intelligent models such as connectionist models is lacking. Hence, this study was undertaken.

2 Materials and Methods

2.1 Data

The sorption data of snack food, *i.e.*, milk protein-rich extruded snack prepared from composite flour, were experimentally generated by Dairy Technology Division at ICAR-National Dairy Research Institute, Karnal, India [2], which have been summarized in Table 1. These data have been utilized for all training and simulation experiments to develop the proposed connectionist models. The experimental values of EMC for respective water activities (*i.e.*, in the range: 0.112–0.971) at 28, 37 and 45 °C for adsorption were considered. Three datasets comprising 24 data points each for three

[1] Soft computing is a consortium of evolving methodologies, which endeavors to exploit tolerance for imprecision, uncertainty, and partial truth to achieve robustness, tractability, and low total cost. It differs from conventional hard computing in the sense that, unlike hard computing, it is strongly based on intuition or subjectivity. Therefore, soft computing provides an attractive opportunity to represent the ambiguity in human thinking with real life uncertainty. Connectionist models, Fuzzy Logic (FL) and Genetic Algorithms (GA) are the core methodologies of soft computing.

different temperatures (*i.e.*, 72 data points in all) were split into two disjoint subsets using the holdout method of cross-validation with 2/3 records (*i.e.*, 16 points) used for training and remaining 1/3 records (*i.e.*, 8 points) for validation of the model.

Table 1. EMC (g water/100 g solids) of milk protein-rich extruded snacks prepared from composite flour at different temperatures (t) and water activities (a_W) for adsorption.

t = 28 °C			t = 37 °C			t = 45 °C		
	EMC			EMC			EMC	
a_W	Mean	SD	a_W	Mean	SD	a_W	Mean	SD
0.112	0.888	0.029	0.107	0.812	0.013	0.103	0.786	0.016
0.230	0.920	0.014	0.212	0.856	0.006	0.198	0.818	0.010
0.326	1.690	0.278	0.316	1.346	0.012	0.308	1.128	0.025
0.441	2.720	0.258	0.435	2.174	0.009	0.430	2.034	0.058
0.530	4.131	0.147	0.512	2.440	0.018	0.497	2.325	0.019
0.756	10.699	0.111	0.740	8.689	0.071	0.726	6.749	0.060
0.845	14.869	0.023	0.827	12.175	0.288	0.817	10.475	0.320
0.971	33.751	1.100	0.966	28.451	0.536	0.961	23.917	0.393

Mean and SD values are based on three replicates.

2.2 Connectionist Models

The connectionist models belong to the branch of Cognitive Science and have originated from diverse sources, ranging from the fascination of mankind with understanding and emulating the human brain, to broader issues of imitating human abilities, such as, speech and the use of language, to the practical commercial, scientific, and engineering disciplines of pattern recognition, modeling and prediction.

The basic structure of a connectionist model consists of processing elements (Fig. 1), and are analogous to biological neurons in the human brain, which are grouped into layers (or slabs). The most common connectionist network structure consists of an input layer, one or more hidden layers and an output layer.

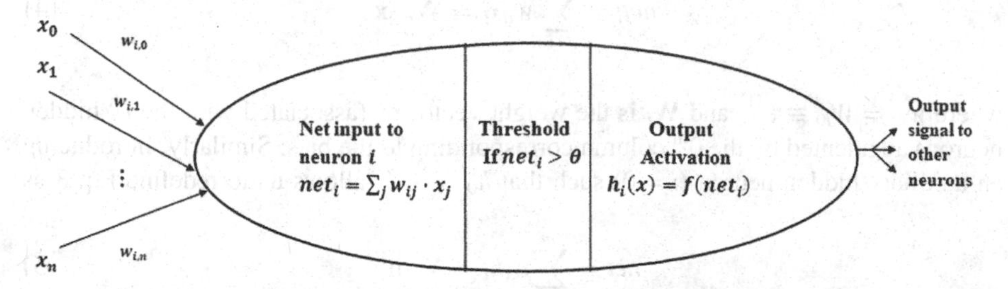

Fig. 1. Schematic representation of general neuron model.

Let the input dimension be $n(n \in Z_+)$ and let the number of hidden neurons be $m(m \in Z_+)$. Z_+ is the set of positive integers. The training pairs are represented by $D = \{\mathbf{x}^{(p)}, t^{(p)}\}$, where $x^{(p)}$ and $t^{(p)}$ denote input and corresponding target patterns; $p = 1, 2, \ldots, P$; $P \in Z_+$, is the number of training exemplars; and the index p is always assumed to be present implicitly. The matrix $\mathbf{w}$ denotes the input to the hidden neurons connection strength, w_{ij} is the $(i, j)^{\text{th}}$ element of the matrix $\mathbf{w}$ representing the connection strength between the jth input and the ith hidden layer neuron. With this nomenclature, the net input to the ith hidden layer neuron is given by

$$net_i = \sum_{j=1}^{n} w_{ij}x_j + \theta_i^{(1)} = \mathbf{w}_i \cdot \mathbf{x} + \theta_i^{(1)} \tag{1}$$

where $\theta_i^{(1)}$ is the bias of the ith hidden layer neuron. The output from the ith hidden layer neuron is given by

$$h_i(\mathbf{x}) = f^{(1)}(net_i) \tag{2}$$

where $f^{(1)}(\cdot)$ is a nonlinear transfer function.

The transfer function determines the output from a summation of the weighted inputs of a neuron. The transfer functions for neurons in the hidden layer are often nonlinear and they provide the nonlinearities for the network. The choice of transfer functions may strongly influence complexity and performance of neural network models. Sigmoidal transfer functions are most commonly used. The net input to the output neuron may be defined similarly as Eq. 1 as follows

$$net = \sum_{i=1}^{m} v_i h_i + \theta^{(2)} = \mathbf{v} \cdot \mathbf{h} + \theta^{(2)} \tag{3}$$

where v_i represents the connection strength between the ith hidden layer neuron and the output neuron, while $\theta^{(2)}$ is the bias of the output neuron. Adding a bias neuron x_0 with input value as +1, Eq. 1 can be rewritten as

$$net_i = \sum_{j=0}^{n} w_{ij}x_j = \mathbf{W}_i \cdot \mathbf{x} \tag{4}$$

where $w_{i0} = W_{i0} \equiv \theta_i^{(1)}$ and $\mathbf{W}_i$ is the weight vector $\mathbf{w}_i$ (associated with the ith hidden neuron) augmented by the 0th column corresponding to the bias. Similarly, introducing an auxiliary hidden neuron ($i = 0$) such that $h_0 = +1$, allows us to redefine Eq. 3 as

$$net = \sum_{i=0}^{m} v_i h_i = \mathbf{V} \cdot \mathbf{h} \tag{5}$$

where $v_0 \equiv \theta^{(2)}$.

The equation for the network output neuron is given by

$$net_o = f^{(2)}(net) = net \tag{6}$$

where $f^{(2)}(\cdot)$ is a linear function.

The notations are diagrammatically exemplified in Fig. 2. This figure represents an n-input, m-hidden neuron and one-output feed-forward connectionist network. Such a connectionist model is trained to fit a dataset D by minimizing an error function (or performance function) as

$$F = E_D(\mathbf{W}) = \frac{1}{P}\sum_{p=1}^{P} \varepsilon^2 = \frac{1}{P}\sum_{p=1}^{P}\left(net_o^{(p)} - t^{(p)}\right)^2 \tag{7}$$

This function is minimized using any standard optimization method.

Error Back Propagation Algorithm

The EBP algorithm *ab initio* is a gradient descent algorithm in which the network weights are moved along the negative of gradient of performance function. The network used is generally of the simple type (Fig. 2), which is called feed-forward network. The term EBP refers to the manner in which the gradient is computed for nonlinear multilayer networks. An EBP network learns by example. We give the algorithm examples of what we want the network to adapt; and it adjusts the network's synaptic weights so that when network is trained substantially, it will produce the required output for a desired input.

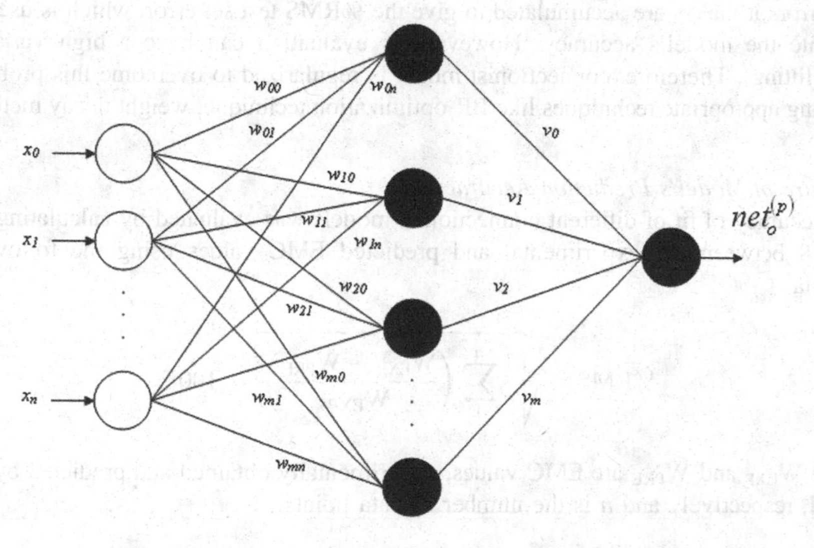

Fig. 2. Schematic of a feed-forward connectionist model.

EBP networks are ideal for simple pattern recognition and mapping tasks. As just mentioned, to train the network, we need to give it examples (input patterns) with corresponding output (called the 'target'). The input and its corresponding target are called a training set. Once the network is substantially trained, it will simulate the desired output for any of the input patterns. That is, the network is first initialized by setting up all its weights to be small random numbers (*e.g.*, between −1 and +1). Next, the input patterns are applied and the output is produced. This is known as forward pass. The network generated output is, generally, different from the actual target, since all the weights are random! We then calculate the error of each neuron, which is essentially: (target − actual output), *i.e.*, 'what we want' minus 'what we actually got'. This error is then used mathematically to adjust the weights in such a manner that the error is minimized. In other words, the output of each neuron will get closer to its target (this part is called the reverse pass or error back propagation). This process is repeated again and again until the error is minimal.

The Neural Network Toolbox under the MATLAB software was used for all simulation experiments. The MATLAB supports several variations of basic EBP algorithm that are based on some heuristics and standard optimization techniques, such as variable learning rate gradient descent, conjugate gradient, Newton methods, *etc*. The EBP learning algorithm based on BR optimization technique has been employed to develop the proposed connectionist model in this study. The 'trial and error' method was used to reach optimum model configuration in all the simulation experiments.

Holdout Method

The holdout method is simplest and commonly used cross-validation technique. The dataset is split into two sets, called the training set and the testing set. The connectionist model is trained using the training set only. Then, the substantially trained connectionist model is applied to simulate the output values for the data kept in the testing set. The errors it makes are accumulated to give the %RMS test set error, which is used to evaluate the model's accuracy. However, its evaluation can have a high variance (over-fitting). Therefore, connectionist model is regularized to overcome this problem by using appropriate techniques like BR optimization technique, weight decay method, *etc*.

Measure of Model's Prediction Accuracy

The accuracy of fit of different connectionist models was evaluated by calculating the %RMS between the experimental and predicted EMC values using the following formula:

$$\%\mathrm{RMS} = \sqrt{\frac{1}{n}\sum_{1}^{n}\left(\frac{\mathrm{W_{EXP}} - \mathrm{W_{PRE}}}{\mathrm{W_{EXP}}}\right)^{2}} \times 100 \tag{8}$$

where $\mathrm{W_{EXP}}$ and $\mathrm{W_{PRE}}$ are EMC values, experimentally obtained and predicted by the model, respectively and n is the number of data points.

3 Results and Discussion

3.1 Intelligent Modelling of Isotherms in Milk Protein-Rich Extruded Snacks

Feed-forward sigmoidal connectionist models based upon EBP learning algorithm with BR mechanism and various combinations of internal parameters such as number of hidden layers, number of neurons in each hidden layer, data splitting scheme, learning rate, regularization constant, error goal, epochs, *etc.*, were empirically investigated to model the adsorption isotherms of milk protein-rich extruded snacks prepared from composite flour, at different temperatures (*i.e.*, 28, 37 and 45 °C) and water activities (*i.e.*, in the range: 0.112–0.971), in this study.

In all simulation experiments, the connectionist models with single hidden layer were found to fit the best to the adsorption data. The number of neurons in the hidden layer was varied between 8 and 12. Tangent-sigmoid and log-sigmoid functions were investigated empirically as transfer functions to be employed on neurons in the hidden layer. While linear function was employed on the output layer neuron. The learning rate, error goal, and epochs were empirically set to 0.01, 0.001 and 100, respectively. The regularization constant was varied between 0.01 and 0.05.

The various configurations of connectionist models explored have been summarized in Table 2. The best models at the three different temperatures have been shown underlined in Table 2. Evidently, the best connectionist models to predict the adsorption isotherms of milk protein-rich extruded snacks prepared from composite flour, at all the three temperatures, *i.e.*, 28, 37 and 45 °C comprised 10 neurons in the hidden layer with tangent-sigmoid transfer function; and attained accuracy as 0.467, 0.958 and 0.636 %RMS, respectively.

The prediction accuracy of connectionist models described above were compared with that of the nine different conventional empirical sorption models (Table 3) fitted to the same dataset to predict the adsorption isotherms of milk protein-rich extruded snacks prepared from composite flour, at three temperatures, *i.e.*, 28, 37 and 45 °C. Apparently, Table 3 revealed that Ferro-Fontan and Peleg were the best similar models among the nine conventional sorption models, with %RMS (underlined) lying in the ranges: 1.63–1.89 and 1.41–3.33, respectively, for the three temperatures, *i.e.*, 28, 37 and 45 °C.

The MSI of experimental data *vis-à-vis* the predicted isotherms produced by the best connectionist models (on test set) of adsorption characteristics in the water activity and temperature ranges studied for milk protein-rich extruded snacks prepared from composite flour are presented in Figs. 3, 4 and 5.

We know that the lower the value of %RMS between predicted and experimental values; the better the goodness of fit. Further, a good description of the isotherm is generally considered to be smaller than 7 %RMS when a model is applied.

Hence, it is deduced that the connectionist sorption models developed in this study seemed to be superior over conventional sorption models fitted to the same dataset, to efficiently and intelligently predict adsorption isotherms of milk protein-rich extruded snacks prepared from composite flour, at different temperatures and over the given range of water activities.

Table 2. Feed-forward sigmoidal connectionist models based on error back propagation algorithm with Bayesian regularization mechanism to predict adsorption isotherms of milk protein-rich extruded snacks prepared from composite flour, at three different temperatures and water activities range.

Number of neurons and transfer function in hidden layer	Transfer function in output layer	Learning rate	Error goal	Regularization constant	Epochs	Data partition		%RMS (on test set)
						Number of records in training set	Number of records in test set	
28 °C								
12 (Tangent-sigmoid)	Linear	0.01	0.001	0.05	100	16	8	0.469
12 (Log-sigmoid)	Linear	0.01	0.001	0.01	100	16	8	1.624
10 (Tangent-sigmoid)	Linear	0.01	0.001	0.05	100	16	8	0.467
10 (Log-sigmoid)	Linear	0.01	0.001	0.01	100	16	8	1.166
8 (Tangent-sigmoid)	Linear	0.01	0.001	0.05	100	16	8	0.470
8 (Log-sigmoid)	Linear	0.01	0.001	0.01	100	16	8	1.372
37 °C								
12 (Tangent-sigmoid)	Linear	0.01	0.001	0.05	100	16	8	0.960
12 (Log-sigmoid)	Linear	0.01	0.001	0.05	100	16	8	1.354
10 (Tangent-sigmoid)	Linear	0.01	0.001	0.05	100	16	8	0.958
10 (Log-sigmoid)	Linear	0.01	0.001	0.01	100	16	8	0.960
8 (Tangent-sigmoid)	Linear	0.01	0.001	0.05	100	16	8	1.180
8 (Log-sigmoid)	Linear	0.01	0.001	0.01	100	16	8	1.008
45 °C								
12 (Tangent-sigmoid)	Linear	0.01	0.001	0.05	100	16	8	0.909
12 (Log-sigmoid)	Linear	0.01	0.001	0.05	100	16	8	0.662
10 (Tangent-sigmoid)	Linear	0.01	0.001	0.05	100	16	8	0.636
10 (Log-sigmoid)	Linear	0.01	0.001	0.05	100	16	8	1.409
8 (Tangent-sigmoid)	Linear	0.01	0.001	0.05	100	16	8	1.195
8 (Log-sigmoid)	Linear	0.01	0.001	0.01	100	16	8	1.016

Table 3. Conventional empirical sorption models to predict adsorption isotherms of milk protein-rich extruded snacks prepared from composite flour, at three different temperatures and water activities range.

Name of the model	%RMS		
	28 °C	37 °C	45 °C
Two-parameter models			
Lewicki-I	7.58	6.09	4.80
Mizrahi	15.74	12.19	10.01
Modified BET	4.69	18.31	20.06
Three-parameter and four-parameter models			
Ferro-Fontan	1.63	1.89	1.752
GAB	1.898	2.485	2.573
Lewicki-II	14.76	12.39	10.49
Modified GAB	3.96	3.66	1.89
Modified Mizrahi	7.874	7.754	6.196
Peleg	3.33	1.68	1.41

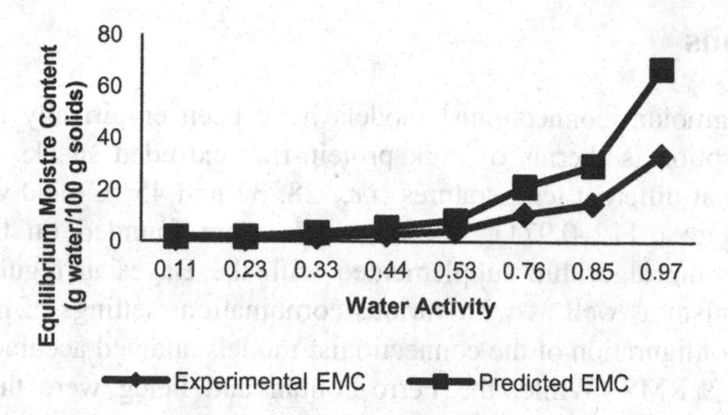

Fig. 3. Equilibrium moisture content predicted by connectionist sorption models *vis-à-vis* experimental data, corresponding to various water activities in the range 0.11–0.97 at 28 °C for adsorption in milk protein-rich extruded snacks prepared from composite flour.

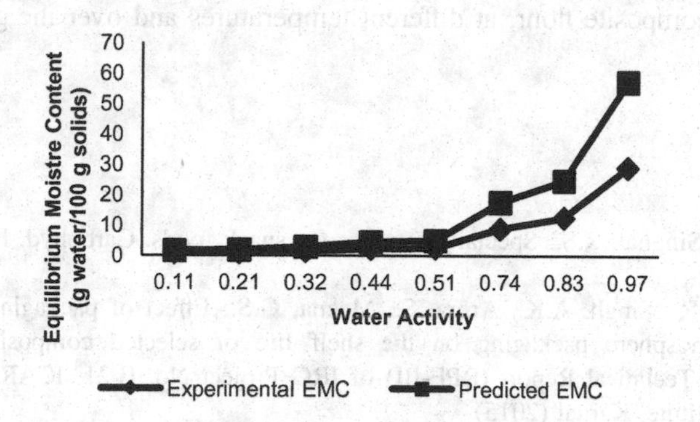

Fig. 4. Equilibrium moisture content predicted by connectionist sorption models *vis-à-vis* experimental data, corresponding to various water activities in the range 0.11–0.97 at 37 °C for adsorption in milk protein-rich extruded snacks prepared from composite flour.

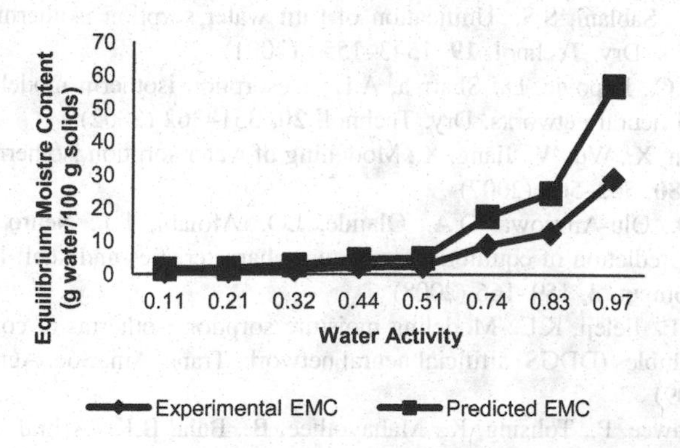

Fig. 5. Equilibrium moisture content predicted by connectionist sorption models *vis-à-vis* experimental data, corresponding to various water activities in the range 0.11–0.97 at 45 °C for adsorption in milk protein-rich extruded snacks prepared from composite flour.

4 Conclusions

Feed-forward sigmoidal connectionist models have been empirically investigated to model the adsorption isotherms of milk protein-rich extruded snacks prepared from composite flour, at different temperatures (*i.e.*, 28, 37 and 45 °C) and water activities (*i.e.*, in the range: 0.112–0.971). These models were founded on the error back propagation learning algorithm supplemented with the Bayesian regularization optimization mechanism as well as with various combinations/settings of internal parameters. The best configuration of the connectionist models attained accuracy in the range of 0.467–0.958 %RMS. While, the Ferro-Fontan and Peleg were the best similar models among the nine conventional sorption models studied in this paper, with % RMS lying in the ranges: 1.63–1.89 and 1.41–3.33, respectively, for the three temperatures, *i.e.*, 28, 37 and 45 °C. Hence, the connectionist sorption models developed in this study were found to be superior over conventional sorption models to efficiently and intelligently predict adsorption isotherms of milk protein-rich extruded snacks prepared from composite flour, at different temperatures and over the given range of water activities.

References

1. Sajilata, M., Singhal, R.S.: Specialty starches for snack foods. Carbohyd. Polym. **59**, 131–151 (2005)
2. Panjagari, N.R., Singh, A.K., Arora, S., Meena, G.S.: Effect of packaging materials and modified atmosphere packaging on the shelf life of selected composite dairy foods. Unpublished Technical Report (RPF-III) of IRC Project No. E-21, ICAR-National Dairy Research Institute, Karnal (2015)
3. Halsey, G.: Physical adsorption on non-uniform surfaces. J. Chem. Phys. **16**, 931–937 (1948)
4. Myhara, R.M., Sablani, S.S., Al-Alawi, S.M., Taylor, M.S.: Water sorption isotherms of dates: modeling using GAB equation and artificial neural network approaches. Food Sci. Technol. **31**, 699–706 (1998)
5. Myhara, R.M., Sablani, S.S.: Unification of fruit water sorption isotherms using artificial neural networks. Dry. Technol. **19**, 1543–1554 (2001)
6. Panchariya, P.C., Popovic, D., Sharma, A.L.: Desorption isotherm modelling of black tea using artificial neural networks. Dry. Technol. **20**, 351–362 (2002)
7. Peng, G., Chen, X., Wu, W., Jiang, X.: Modelling of water sorption isotherm for corn starch. J. Food Eng. **80**, 562–567 (2007)
8. Araromi, D.O., Olu-Arotiowa, O.A., Olajide, J.O., Afolabi, T.J.: Neuro fuzzy modelling approach for prediction of equilibrium moisture characteristics and shelf life of corn flour. Int. J. Soft Comput. **3**, 159–166 (2008)
9. Kingsly, A.R.P., Ileleji, K.E.: Modeling moisture sorption isotherms of corn dried distillers grains with solubles (DDGS) artificial neural network. Trans. Am. Soc. Agric. Biol. Eng. **52**, 213–222 (2009)
10. Janjai, S., Intawee, P., Tohsing, K., Mahayothee, B., Bala, B.K., Ashraf, M.A., Müller, J.: Neural network modelling of sorption isotherms of longan (*Dimocarpus longan* Lour.). Comput. Electron. Agric. **66**, 209–214 (2009)

11. Chayjan, R.A.: Modeling of sesame seed dehydration energy requirements by a soft-computing approach. Aust. J. Crop Sci. **4**, 180–184 (2010)
12. Chayjan, R.A., Esna-Ashari, M.: Comparison between artificial neural networks and mathematical models for estimating equilibrium moisture content in raisin. Agric. Eng. Int. CIGR J. **12**, 158–166 (2010)
13. Correa, P.C., Goneli, A.L.D., Junior, P.C.A., de Oliveira, G.H.H., Valente, D.S.M.: Moisture sorption isotherms and isosteric heat of sorption of coffee in different processing levels. Int. J. Food Sci. Technol. **45**, 2016–2022 (2010)
14. Chayjan, R.A., Esna-Ashari, M.: Effect of moisture content on thermodynamic characteristics of grape: mathematical and artificial neural network modelling. Czech J. Food Sci. **29**, 250–259 (2011)
15. Gazor, H.R., Eyvani, A.: Adsorption isotherms for red onion slices using empirical and neural network models. Int. J. Food Eng. **7** (2011). https://doi.org/10.2202/1556-3758.2592
16. Morse, G., Jones, R., Thibault, J., Tezel, F.H.: Neural network modelling of adsorption isotherms. Adsorption **17**, 303–309 (2011)
17. Sharma, A.K., Lal, M., Sawhney, I.K.: Computational aspects of soft computing models to predict sorption isotherms in Nutrimix (weaning food). Math. Eng. Sci. Aerosp. **5**, 105–119 (2014a)
18. Sharma, A.K., Sawhney, I.K., Lal, M.: Intelligent modelling and analysis of moisture sorption isotherms in milk and pearl millet based weaning food 'fortified Nutrimix'. Dry. Technol. **32**, 728–741 (2014b)
19. Sharma, A.K., Sawhney, I.K.: Modelling moisture sorption characteristics in dried acid casein using connectionist paradigm vis-à-vis classical methods. J. Food Sci. Technol. **52**, 151–160 (2015)

AI Based HealthCare Platform for Real Time, Predictive and Prescriptive Analytics

Jagreet Kaur[1(✉)] and Kulwinder Singh Mann[2]

[1] IKGPTU, Jalandhar, India
jagreet@xenonstack.com
[2] IT Department, GNDEC, Ludhiana, India

Abstract. The healthcare industry is changing at a fast rate. Recently, big data real time computing has been studied to enhance the quality of healthcare services and reduce costs making decisions in real-time. Artificial intelligence is used to track the big data. AI in healthcare sector could make treatment plans better, and also provide physicians information they need to make a good decision. This paper proposes a generic architecture for big data healthcare analytic by using open sources, including Apache Spark, Apache NiFi, Kafka, Tachyon, Gluster FS, Elastic search and NoSQL Cassandra. This paper will show the importance of applying AI based predictive and prescriptive analytics techniques in Health sector. The system will be able to extract useful knowledge that helps in decision making and medical monitoring in real-time through an intelligent process analysis and big data processing.

Keywords: Big data · Elastic search · Health care analytics · Kafka NiFi · NoSQL cassandra · Real-time · Spark · Stream computing

1 Introduction

As with the change in the time and advancements in the technology, there is a need to make a systematic change to health systems to improve the quality, efficiency, and effectiveness of patient care. Chronic diseases like heart disease, stroke, cancer, and diabetes are considered as the most expensive, common and preventable health problems but due to the poor health care systems, patients can't able to take good care for the problems.

The strategic aim of value based Health Care is to ensure that everyone can use the health services they are needed for their good health and well-being. These services range from clinical care to the public services for individual patients that are helpful for the health of whole populations.

There is a need to improve the healthcare quality and coordination so that outcomes are consistent with current professional knowledge.

J. Kaur—Research Scholar.

R. Sharma et al. (Eds.): ICAN 2017, CCIS 805, pp. 138–149, 2018.
https://doi.org/10.1007/978-981-13-0755-3_11

- The cost of treatment for the problems should be reduced.
- Need to go in depth to find the origin of the disease.
- Medical errors by the doctors should be reduced and improve the patient safety
- Need to detect the spreading diseases like viral, dengue at the earlier stage.
- Need to improve the treatment methods.

Health care involves a diverse set of public and private data collection systems having different sources of Data. According to the Health Data, most of the information is not in structured, relational format. As shown in Fig. 1. Text or natural language data resides in many fields including several fields such as physician notes, nursing notes, surgical notes, radiology notes, pathology reports, admission notes, clinical data, Genomic data, Behaviour data etc. These fields may have valuable information about the patient including diagnosis, history, family history, complaints, statistics, and opinions.

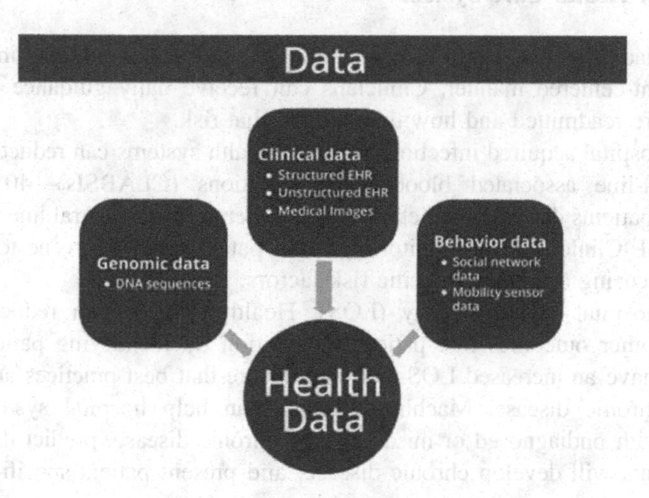

Fig. 1. Health care data sources

It is very difficult to use this massive amount of data from different sources to take the right decision for the right patient at the right time by the doctors. This will slow down the personalised care to the patient. So, it is necessary to develop a new strategy or system to care for patients which involves the health of patients while decreasing the cost of care.

The objective of this paper is to prove that healthcare data analytics techniques are not useful enough and suitable these days to manage big data issues and improves healthcare data analytics due to the rapid growth and evolution of technology. It also aims to promise experts of a enhance quality of medical results, as well as lower time needed to analyze healthcare data by keeping organizations up to-date and sort medical data in a logical manner along with access and retrieve patient's historical data fast and smooth and efficiently.

The rest of this paper is organised as follows:

Section 2 gives an overview about use of cloud computing in field of Health care sector as well as discussed, the challenges of big data and overview of knowledge discovery.
Section 3 describes the Big data components and 7 v's of Big data.
Section 4 describes the related work
Section 5 describes the platform design in which steps needed to do analytics will be discussed.
Section 6 describes the architectural design
Section 7 gives an overview about artificial intelligence that helps the system to understand from its experience as well as use cases of AI based Healthcare System Architecture.

1.1 Need of Health Care System

- Reduce readmissions. Health care systems can reduce readmissions in an efficient and patient-centered manner. Clinicians can receive daily guidance as to which patients are readmitted and how they reduce that risk.
- Prevent hospital acquired infections (HAIs). Health systems can reduce HAIs, such as central-line associated bloodstream infections (CLABSIs)—40 percent of CLABSI patients die—by predicting which patients with a central line will develop a CLABSI. Clinicians can monitor high risk patients and intervene to reduce that risk by focusing on patient-specific risk factors.
- Reduce hospital Length-of-Stay (LOS). Health systems can reduce LOS and improve other outcomes like patient satisfaction by identifying patients that are likely to have an increased LOS and then ensure that best practices are followed.
- Predict chronic disease. Machine learning can help hospital systems identify patients with undiagnosed or misdiagnosed chronic disease, predict the likelihood that patients will develop chronic disease, and present patient-specific prevention interventions.
- Reduce 1-year mortality. Health systems can reduce 1-year mortality rates by predicting the likelihood of death within one year of discharge and then match patients with appropriate interventions, care providers, and support.
- Predict propensity-to-pay. Health systems can determine who needs reminders, who needs financial assistance, and how the likelihood of payment changes over time and after events.
- Predict no-shows. Health systems can create accurate predictive models to assess, with each scheduled appointment, the risk of a no-show, ultimately improving patient care and the efficient use of resources [1].

2 Cloud Computing Enabled Applications

Cloud computing is a custom of using a network of remote servers accommodated on the Internet to manage, store and process data, rather than a personal computer.

Healthcare is moving towards digital platform, more patient-oriented and data-driven. Availability of data, irrespective of the location of the patient and the clinician, has become the key to both patient satisfaction and improved clinical outcomes. Cloud technologies can significantly facilitate this trend.

Healthcare data needs to be shared across various settings and geographies which further burden the healthcare provider and the patient causing significant delay in treatment and loss of time. Cloud caters to all these requirements thus providing the healthcare organizations an incredible opportunity to improve services to their customers, the patients, to share information more easily than ever before, and improve operational efficiency at the same time.

2.1 Big Challenges in Health Care System

- No unified format: As there is no standard method adopted in hospitals, labs to generate reports etc., they generate reports in their own formats so it is a challenge to integrate different formats and process the data.
- Unstructured data: Need to understand unstructured clinical notes in the right context. If we take the case of doctors, we can even not able to understand the handwriting of the doctors as there is no standard format used by the doctor while providing prescriptions to the patients and hence it is a big issue to process this data.
- Large volume of data: If we take the case of image data like X-rays etc., large size of data is generated for a single patient and there is a need to efficiently handle such data and extract useful information.
- Real time data handling: Need to capture the patient's behavioral data through several sensors, social interactions and communications [2].

2.2 Real-Time Analytics and Predictive Analytics: - Knowledge Discovery

To take the Right decision for the Right patient at the right time by the doctors, Doctors/Physicians need knowledgeable Data, Real-Time Analytics, and Predictive Modelling to improve the health care [3].

Real time monitoring techniques makes data up to date and increase the quality of information by real time streaming of data.

Knowledgeable data is basically the knowledge extracted from information. It is essential for healthcare organizations to effectively manage both internal knowledge and externally generated knowledge to provide the best possible health care facilities to the patients. It increases decision-making capabilities of Physicians.

Predictive analytics involves a variety of statistical techniques from predictive modelling, machine learning and data mining that analyze current and historical facts to make predictions about future. Predictive Modelling benefit to large Healthcare providers to discover the treatment for personalized care managing large patient populations. It helps the physicians to predict the problem of the patient.

Predictive analytics can help to avoid and moderate inaccurate prediction costs along with time that it makes the data sourcing cost lower by specifying the desired and useful data only, since the data is simple, standardized and present in historical clinical databases.

2.2.1 Methods to Predict the Problems

- Regression: Physicians can predict whether the patient is suffering from disease or not by using a regression method.
- Classification: Physicians can easily classify the number of patients suffering from problem. Also diagnose the patient's problem by finding the group in which other patient having same problem lies.
- Comparison: By comparing the patient's symptoms with the symptoms of other patients stored in knowledge dataset and prescribe the proper medication.
- Real-time healthcare analytics can help to improve the quality of care, cost, and meet various requirements by streamlining and automating the process of collecting and measuring vast amounts of healthcare data.

2.2.2 Real-Time Analytics Is Hard

- Can't Stay Ahead. Need to handle multi-sourced and multi-typed of data, including unstructured and semi-structured data. Relational databases like SQL aren't capable of handle this.
- Can't Scale. Need to analyze terabytes or petabytes of data having response times of sub seconds. Single server can't handle this.
- Batch. Batch processes are needed for some jobs. But many times, we need to analyze rapidly changing, multi-structured data in real time. It is very difficult to do ETL processes to cleanse data at that time.

3 Big Data in Terms of Healthcare Sector

3.1 Big Data Components

- Structured Data- Small scale personal Health records, Insurance data
- Unstructured data- Treatment data, Research data, Procedures.
- Sensor data-Medical research data, Telehealth
- New data types-National Electronic Health records, Scan/images, videos

3.2 Big Data Is Categorized as 7 V's

- Volume: Each single X-ray of patient generates a lot of data.
- Velocity: Large number of patients are treated at time.

- Variety: Different data sources generated different type of data.
- Veracity: Accuracy of data is needed.
- Variability: Dynamic behavior of data i.e. changed with time.
- Value: Useful data.
- Visualization: After processing, need a way to present the data in a manner that's readable and accessible.

4 Related Works

Nowadays, Healthcare analytics is a new trend in the field of analytics. Due to the advancements in technologies and science in Health care sector, there is a major breakthrough in data collection. To improve the processing of predictable healthcare system, we have proposed a series of Big Data Health Care System. There are many techniques proposed in order to proficiently process large volume of medical record.

Akshay Raul, Atharva Patil, Prem Raheja, Rupali Sawant proposed "Knowledge Discovery, Analysis and Prediction in Healthcare using Data Mining and Analytics" [4]. In this paper, Author proposed a Healthcare system that will be used to create a public awareness about alternative drugs for a specific medicine, availability of that alternative medicine in an area. The Proposed system helps the patients to find an alternate medicine which is prescribed by the doctor.

Aditi Bansal and Priyanka Ghare proposed "Healthcare Data Analysis using Dynamic Slot Allocation in Hadoop". In this paper HealthCare System is analysis using Hadoop using Dynamic Hadoop Slot Allocation (DHSA) method. This paper proposed a framework which focus on improving the performance of MapReduce workloads and maintain the system [5]. DHSA will focuses on the maximum utilization of slots by allocating map (or reduce) slots to map and reduce tasks dynamically.

Van-Dai Ta, Chuan-Ming Liu, Goodwill Wandile Nkabinde proposed "Big Data Stream Computing in Healthcare Real-Time Analytics". In this paper, Author proposed a generic architecture for big data healthcare analytic by using open sources, including Hadoop, Apache Storm, Kafka and NoSQL Cassandra [6].

Wullianallur Raghupathi and Viju Raghupathi has proposed "Big data analytics in healthcare: promise and Potential". In this paper author proposed the potential of big data analytics in healthcare [7]. The paper provides an overview of big data analytics for healthcare practitioners and researchers. Big data analytics in healthcare is growing into a promising field for providing vision from very large data sets and improving results while reducing costs. This paper proposes a generic architecture for big data healthcare analytic by using open sources, including Hadoop, Apache Storm, Kafka and NoSQL Cassandra.

However, all these works either consider the techniques to extract features form specific healthcare data sources or only focus on batch-oriented task to compute which has higher latency computing. In our proposed framework, data sources are supposed constantly coming with high rate, variety of formats, and high volume. Stream computing of real data and use of cloud computing in serving layer will enhance the results of healthcare analytics.

5 Platform Design

This demonstrates the major components of the Healthcare Analytical platform and the types of services that is needed to build a model.

Data Ingest: The first problem that everyone gets into is data ingest. The device data is coming from anywhere and of any type, it will be represented in a standard manner to analyze.

In this part of the application, solution of these problems is discussed:

- How to ingest data from medical devices?
- How to transform the device data into a format that can be analyzed in a Streams application?
- What are the common data schema types when analyzing medical device data?

Data Preparation: Next, there is a need to prepare the data to analyze. Some of the common problems need to remove include Deduplication, Resampling, Normalization, Cleaning, Noise reduction etc.

Data Processing: Data processing is of two types. Base Analytics is a platform which provide a set of basic analytics of data. These analytics can be used as building blocks for the complex analytics and prediction models. Here are some of the things we can do like Simple vital analytics (calculating rolling average, raising alerts when vitals going beyond normal range), Analytics of ECG, EEG, ICP Waveforms etc. Aggregated Analytics is an area where we combine and aggregate results from the base analytics to form more sophisticated analytic rules [8]. For example, for septic shock detection, the user should be able to describe a rule like this:

- if temperature is >37 °C or <33 °C
- and if heart rate is >90
- and respiratory rate is >19 or PaCO2 < 31 mm HG
- and WBC > 13,000/mm^3, or <5000/mm^3, or >9% the heart
- then raise an alert for early septic shock detection.

Patient Data Correlation/EMR Integration: For more complex analytics, we may need to use EMR data or doctor's notes. For example, we may need to retrieve patient's medical history. Or need to retrieve doctor's notes that include some of the doctor's observations.

These types of data can be ingested from existing hospital infrastructure using the HL7/FHIR protocol.

Central Monitoring Dashboard: As part of the platform, there is a need to create a simple dashboard. This will help users to visualize their data and validate their analytic results [9]. The dashboard can be web-based, mobile.

Alert and Notification Framework: When an important event occurs, we need to be able to notify and alert the right people to help the patient. In this part of the framework, we need to deliver notification to the right people based on alert types and patient information, alerts can be delivered via email, text messaging, etc. or should be displayed onto a dashboard.

6 Healthcare Platform Architecture

In this section, we propose an architecture including the advantages of batch and stream computing to enhance big data computing in healthcare [9]. It can also deal with a large data set by providing cost reduction along low latency and better health care conditions.

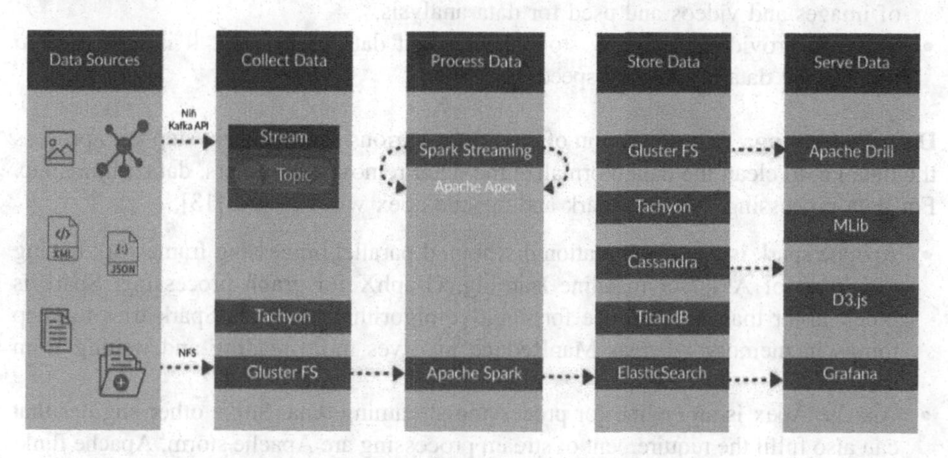

Fig. 2. Data pipelining

The Fig. 2 shows the detailed architecture of Healthcare Platform.

Data Sources: As there are diverse sources of Health data available now a day like data from medical devices, Social media, EMR, Back- office-systems, Pharmacy, Medical images and HL7 events [10].

Data Ingestion: While developing the system, the first step needed is to integrate the data i.e. collect the data from various data sources. Data Ingestion can be from multiple sources, so need a unified platform where we can manage all our data sources. Here the data can be generated from two types of sources, one is real time data and other is batch data.

To ingest data from real time data sources, Apache NiFi and Kafka API can be used.

- Apache NiFi- Apache NiFi is distributed, scalable, fault-tolerant workflow automation platform. So, we will be having unified view on our Web UI from where we can monitor all our different data sources from where data ingestion is taking place.
- Apache Kafka - Apache Kafka is a Distributed Messaging System and disk based cache. It can store the data that is collected by Apache NiFi. It provides unified, high-throughput, low latency and works on publish-subscribe model.

To ingest data in batches, NFS (Network File system) can be used [11]. Data from Batch files can be ingested using Apache NiFi and Kafka API also but to process older data that was stored on distributed servers can only be accessed through NFS. NFS protocol provides remote access to shared disks across networks. An NFS-enabled server can share directories and files with clients, allowing users and programs to access files on remote systems as if they were stored locally. An NFS-mounted cluster allows easy data ingestion of data sources such as files, images from other machines leveraging standard Linux commands, utilities, applications, and scripts [12].

- Gluster FS is free and open source software that can be used for distributed storage of images and videos and used for data analysis.
- Tachyon provide in memory storage with fast data processing. It is also used to process the data at memory speed.

Data Processing: After ingestion of data from various sources, next step is to process the data i.e. to clean the data, normalize the data, remove all outliers, data mining, etc. For data processing, Apache spark and apache apex will be used [13].

- Apache spark is a next generation distributed parallel processing framework having a rich set of APIs for machine learning, GraphX for graph processing. Spark is much faster than MapReduce for iterative algorithms because Spark tries to keep things in memory whereas MapReduce involves more reading and writing from disk.
- Apache Apex is an engine for processing streaming data. Some other engines that can also fulfil the requirement of stream processing are Apache storm, Apache flink. But Apache Apex has a built-in support for fault-tolerance, scalability and focus on operability [14].
- Apache Flume is a distributed, reliable, and available service for efficiently collecting, aggregating, and moving large amounts of log data. It has a simple and flexible architecture based on streaming data flows. It is robust and faults tolerant with reliability mechanisms and many failovers and recovery mechanisms.

Data Storage: As already discussed, Health care data consists of a variety of data like images, doctor notes, lab reports, insurance data etc. Different types of storage system are also needed like cloud to store the data. To store the data Gluster FS, Tachyon, Cassandra, Titan DB and Elasticsearch can be used.

- Gluster FS is free and open source software that can be used for distributed storage of images and videos and data analysis.
- Tachyon is used to process the data at memory speed. Also provide in memory storage with fast data processing.
- Apache Cassandra is a free and open-source distributed database management system used to handle large amounts of data like sensor data and claim data. It is an open source project developed by Facebook
- Titan dB is a database used to store graph and link analysis.
- Elasticsearch is a search engine used for indexing.

Data Visualization: Last step is to visualize the data. Now the stored data can be used for any type of querying, means present the data and for presentation, Apache drill, MLib, D3.js and grafana can be used.

- D3.js is a JavaScript library for visualization of data.
- MLib is a machine learning library using which various machine learning algorithms can be implemented for predictive analysis of data.
- Grafana is an open source visualization tool and graph editor for Graphite.
- Drill is a querying language that can read from all kinds of data, process petabytes of data and trillions of records in seconds and can be used by data analysts along with tools like Tableau for fast visualizations.

7 AI Based Healthcare System

AI refers to 'Artificial Intelligence' which means making machines capable of performing intelligent tasks like human beings. AI performs automated tasks using intelligence.

There is need to make system AI enabled so that machine can automatically predict and prescribe the results from its own experience [15]. To make system AI enabled, Natural Language processing, Knowledge representation, Automated reasoning, and Machine Learning should be in the system as shown in Fig. 3. NLP helps to make computer/machines as intelligent as human beings in understanding language.

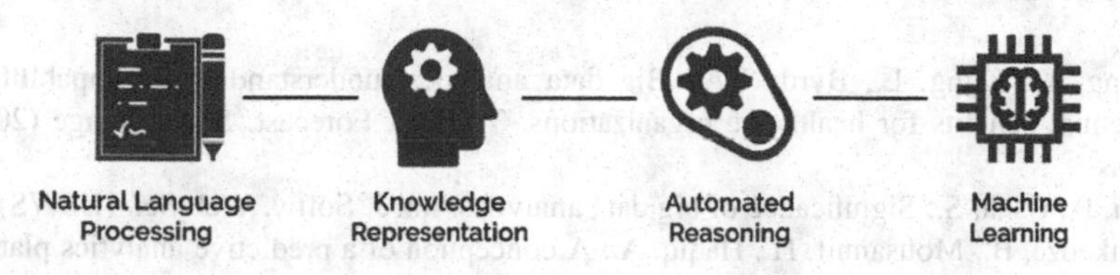

Fig. 3. AI enabled healthcare system

7.1 Use Cases of AI Based Healthcare System Architecture

One of the most effective use of AI based Healthcare system is to optimize the clinical process. AI based mobile app ask patients about their symptoms and provides an easy to understand information about their Health. The system uses natural Language processing to provide a rich experience and machine learning algorithms to create a map of patient condition and give a personalized experience.

The treatment and prevention of dangerous diseases depends on detecting the symptoms at the right time. In various cases, early diagnosis can result in complete cure. Inversely, a late or wrong diagnosis can have damaging results. It is difficult for humans to make reliable decisions.

AI algorithms can rapidly ingest billions of samples in short order and collect useful patterns. Machine learning algorithms can be used to make the knowledge bases used by expert systems and predictive analysis of data. Hence, AI based Healthcare system is helpful for patients as well as doctors.

8 Conclusions

This paper presents big data computing in healthcare applications. The challenges in big data analysis can be understood by 7 V. The proposed architecture can support for healthcare analytics by providing batch and stream computing, extendable storage solution and query management. To achieve more efficient result from healthcare, there is a more need to handle the health care data that is growing time by time at high rate, and larger scale with tons of inconsistent data sources. A distributed system should be arranged to interchange data among labs, hospital systems, and clinical centers. Previous analytics tools have become transparent and friendly. But, when they emerged with open source development tools and platforms, it will become very complex, need complex programming and need the application of a variety of skills. Additionally, Tools for data analytics such as machine learning, Tableau public, Open Refine, KNIME and data mining can be used to improve the efficiency of data analytics.

Acknowledgement. We express our sincere thanks to Mr. Navdeep Singh Gill (C.E.O of XenonStack Pvt. Ltd. and Founder of Akira.ai) for providing us the opportunity to test the proposed architecture in Real time data.

References

1. Wang, Y., Kung, L., Byrd, T.A.: Big data analytics: understanding its capabilities and potential benefits for healthcare organizations. Technol. Forecast. Soc. Change (2016). In press
2. Jain, P., Osha, S.: Significance of big data analytics. Int. J. Softw. Web Sci. (IJSWS) (2015)
3. Boukenze, B., Mousannif, H., Haqiq, A.: A conception of a predictive analytics platform in healthcare sector by using data mining techniques and Hadoop. Proc. Conf. Int. J. Adv. Res. Comput. Sci. Softw. Eng. **6**(8) (2016)
4. Raul, A., Patil, A., Raheja, P., Sawant, R.: Knowledge discovery, analysis and prediction in healthcare using data mining and analytics. In: Proceedings of Conference, 2nd International Conference on Next Generation Computing Technologies (NGCT-2016), Dehradun, India, 14–16 October 2016 (2016)
5. Bansal, A., Ghare, P.: Healthcare data analysis using dynamic slot allocation in Hadoop. Proc. Conf. Int. J. Recent Technol. Eng. (IJRTE) **3**(5) (2014). ISSN: 2277-3878
6. Ta, V.-D., Liu, C.-M., Goodwill Wandile Nkabinde: Big data stream computing in healthcare real-time analytics. In: Proceedings of Conference International Conference on Cloud Computing and Big Data Analysis. IEEE (2016)
7. Raghupathi, W., Raghupathi, V.: Big data analytics in health care: promise and potential. Health Inf. Sci. Syst. **2**(1), 3 (2014)
8. Roopa, M., Manju Priya, S.: A review of big data analytics in healthcare. Proc. Conf. Int. J. Sci. Res. Dev. Sp. Issue – Data Min. (2015)
9. Borana, M., Giri, M., Kamble, S.: Healthcare data analysis using Hadoop. Proc. Conf. Int. Res. J. Eng. Technol. (IRJET) **02**(07), 583 (2015)
10. Sethy, R., Panda, M.: Big data analysis using Hadoop: a survey. Proc. Conf. Int. J. Adv. Res. Comput. Sci. Softw. Eng. **5**(7) (2015)

11. Archenaa, J., Mary Anita, E.A.: A survey of big data analytics in healthcare and Government. In: Proceedings of 2nd International Symposium on Big Data and Cloud Computing (ISBCC 2015), pp. 408–413 (2015)
12. AbdulAmeer, D.A.H.: Medical data mining: health care knowledge discovery framework based on clinical big data analysis. Proc. Conf. Int. J. Sci. Res. Publ. **5**(7) (2015)
13. da Silva Morais, T.: Survey on frameworks for distributed computing: Hadoop, spark and storm. In: Proceedings of Conference Doctoral Symposium in Informatics Engineering (2015)
14. Pathak, H., Rathi, M.: Introduction to real-time processing in Apache Apex. Proc. Conf. Int. J. Res. Advent Technology (2016)
15. Forkan, A.R.M., Khalil, I.: Big data for context-aware monitoring – a personalized knowledge discovery framework for assisted healthcare. IEEE Trans. Cloud Comput. (2015)

Networks

Model Predictive Optimization for Energy Storage-Based Smart Grids

Tsai-Chen Yang and Pao-Ann Hsiung(✉)

National Chung Cheng University, Chiayi, Taiwan
pahsiung@cs.ccu.edu.tw

Abstract. In recent years, energy storage systems (ESS) have started to play the role of an active electricity supplier so as to minimize overall electricity costs in a smart grid. However, ESS lifetime decreases with each cycle of charge/discharge. There is a tradeoff between ESS lifetime and electricity cost saving. As a solution, this work proposes a *Model Predictive Optimization* (MPO) method for distribution management in smart grids. Future energy states are predicted using an *Autoregressive Integrated Moving Average* (ARIMA) model. Based on the predicted electricity status, a near-optimal schedule for ESS usage is found using Genetic Algorithm such that a tradeoff is made between cost saving from electricity trading and the loss of life (LoL) in ESS. Experiment results show that the error rate of the prediction model is less than 10%. The MPO method achieves an overall cost saving of 0.85% and an ESS LoL reduction of 12.18%.

1 Introduction

According to the Tracking Clean Energy Progress 2015 report [1] from the International Energy Agency (IEA), the global annual increase in electricity demand has grown at a rate of 3.6% from 2002 to 2012. With the rapid growth of electricity demand, people try to explore various renewable energy resources. To increase the reliability of energy supply, the scalability of energy management, and the flexibility of energy utilization, the architecture of *smart grids* [5] has been proposed. The smart grid is a distributed system and is composed of multiple micro-grids. Each micro-grid includes multiple renewable power generators (e.g., photovoltaic, wind turbines), the energy storage system (ESS) and power consumers.

Although using renewable energy resources can help reduce the cost of electricity, there is a problem of intermittency in power generation by renewable energy resources. The intermittent problem of renewable energy resources leads to decrease in quality of electricity supply. As a remedy, ESS can be used to improve the power quality [18]. When there is surplus electricity in a micro-grid, it can be used to charge ESS and/or it can be sold to the utility or to other micro-grids. On the contrary, when there is deficient electricity in a micro-grid, ESS can be discharged to cover the deficit and/or electricity can be bought from

R. Sharma et al. (Eds.): ICAN 2017, CCIS 805, pp. 153–165, 2018.
https://doi.org/10.1007/978-981-13-0755-3_12

the utility or from other micro-grids. However, the rate of charging/discharging, over-charging and over-discharging shorten the lifetime of ESS [11].

To find an optimal ESS schedule with a trade-off between its lifetime and cost saving for consumers, we propose an energy storage distributed management system based on *Model Predictive Control* (MPC) technique [8]. The proposed system includes a prediction model and an optimizer. The prediction model predicts the future electricity supply and demand with the time series analysis for history data of the power generators and power loads in micro-grids. According to the predicted results, the optimizer will not only satisfy the requirement with electricity trading, but also effectively choose the appropriate time points and the amount of energy to charge/discharge ESS.

This work is organized as follows. In Sect. 2, we introduce the related work on smart grid. Section 3 gives the overall framework and the details of proposed distribution management method. The experimental results and test cases will be explained and compared in Sect. 4. In Sect. 5, we give the conclusions for this work.

2 Related Work

Since the 20th century, MPC method which incorporates both forecasts and newly updated information has been used in the industrial process [19]. In recent years, MPC has drawn the attention of power systems also [16,17]. The core concept in MPC [3] is to forecast the development in the future timeslots, and to implement control actions in the next timeslot only. MPC is a rolling process that runs the results from optimizer repeatedly with updated forecasts. With the state feedback from the plant, the prediction model will be adjusted.

According to the stability or the instability of history data, the prediction methods can be classified into the linear method and non-linear method. However, the linear prediction method cannot deal with complicated data such as generation of solar power and wind speed. They are both linear, as well as, non-linear data. Thus, a prediction model which can handle both linear and non-linear data is more often to be used. For example, Box et al. proposed an Autoregressive Integrated Moving Average (ARIMA) model [2] for time series. The ARIMA model can reduce the prediction error because of the precise noise model. With the differences added to the mode, a non-stationary series can be transformed into a stationary series. The ARIMA model has the capability to handle many kinds of prediction questions. As a result, the ARIMA model was introduced to predict linear and non-linear data, simultaneously.

Optimization in smart grid enables consumers to use power related data to make more informed and profitable decisions such as power distribution and cost management [6]. Therefore, over the past few decades, a large number of research has focused on solving optimization problems. For example, *Simulated Annealing* (SA) [22] algorithm and *Tabu Search* (TS) [7] algorithm. However, both of them are not good at solving large-scale problems because of the slow convergence and long execution time. In the recent years, *Genetic Algorithm*

(GA) [10] is frequently applied to optimize parameter configurations for different tasks in several years. For example, Yildirim et al. used GA to minimize energy consumption [24], Lee et al. used GA to determine the power consumption scheduling [12], and Chung et al. used GA to decide production frequencies of economic lot scheduling [4]. Miao et al. also used GA to minimize the entire energy expense with the power consumption scheduling [15].

To solve the problem of intermittency generated by renewable energy sources, ESS has played an important role [21]. ESS improves the stability of the electricity usage. It helps perform ancillary services such as peak-shaving and outage protection. As a service in peak-shaving, ESS can determine whether to charge or discharge for cost saving in peak time of electricity. As a service in outage protection, ESS can be an electricity supplier in which case it is used as a backup. However, the higher usage frequency of the battery module in ESS, the greater will the battery lifetime decrease. Therefore, a trade-off between the cost saving of electricity and lifetime of batteries should be considered.

3 Distribution Management System Design

In this section, the proposed design of distribution management system is described. We introduce the core techniques used in this system including MPC, the prediction model, and the optimization algorithm.

3.1 Model Predictive Optimization

Figure 1 shows the architecture system of the proposed Model Predictive Optimization (MPO) method for smart grids. Each micro-grid is equipped with the capabilities of predicting power generation and demand loads. A micro-grid not only satisfies demand loads via the utility, but also through other micro-grids via trading so as to save costs and earn additional economic benefits.

The distribution management system includes two main parts, namely prediction models and an optimizer. Each micro-grid has its own prediction model for predicting demand loads and renewable energy generation. Through collecting history demand loads and renewable energy generation data, a micro-grid can forecast the situation of electricity such as deficient or surplus. Then, all micro-grids send the prediction information including forecasted demand loads and forecasted renewable energy generation to the optimizer, which is basically an ESS scheduler in this work.

When the optimizer receives the prediction information of demand loads and generation, it will use an optimization algorithm to schedule future control inputs like trading electricity capacity such as buying deficient or selling surplus electricity capacity from the utility or other MGs, and charging/discharging ESS capacity. According to the control inputs, the distribution equipment controller tries to satisfy the electricity requirements. Feedback on the actual situations of electricity demand-response are given to the prediction models. If there is a significant difference between actual data and predicted data, the prediction is rectified (re-trained) based on the feedback.

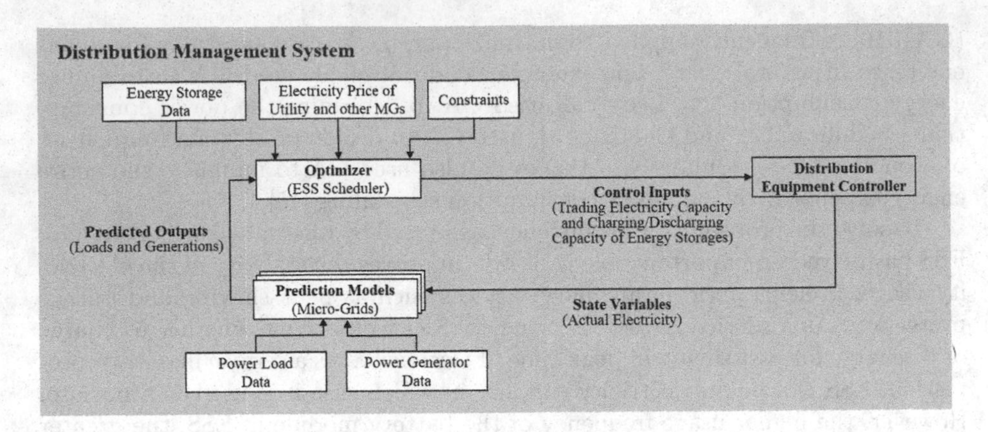

Fig. 1. Architecture of model predictive optimization-based distribution management system

3.2 Prediction with ARIMA Model

The inputs of proposed prediction model include history electricity data of power loads and renewable energy resources. We use the ARIMA model proposed in the Box-Jenkins methodology [2] to forecast the demand loads and power generation in the future. The ARIMA *(p,d,q)* model is defined in Eq. 1.

$$x_t = \varphi_1 x_{t-1} + \cdots + \varphi_p x_{t-p} + \varepsilon_t - \theta_1 \varepsilon_{t-1} + \cdots + \theta_q \varepsilon_{t-q} \tag{1}$$

The forecasted data x_t is calculated by the autoregressive parameters from φ_1 to φ_p, the history series data from x_{t-1} to x_{t-p}, the moving average parameters from θ_1 to θ_q, and a series of random errors (or residuals) from ε_t to ε_{t-q}. This model observes the regression on history data x_{t-i} for p periods and has a series of random errors ε_{t-i} for q periods. The two values p and q are selected by observing the *autocorrelation function* (ACF) and *partial autocorrelation function* (PACF). However, since the ARIMA model is only for stationary series, we have to make differencing of the non-stationary series. Let the order of differencing be demoted as d.

After the identification of ARIMA *(p,d,q)* model, we would use the *maximumlikelihood estimation* (MLE) to estimate the parameters of the model. We use the *Akaike Information Criterion* (AIC) [9] and the *Bayesian Information Criterion* (BIC) [20] to check for an adequate model. To get the best fitness of ARIMA model, we should find the minimum values of AIC and BIC as follows:

$$AIC = 2k + n \ln(RSS/n) \tag{2}$$

$$BIC = k \ln(n) + n \ln(RSS/n) \tag{3}$$

where n is the sample size, k is the number of estimated parameters, and RSS is the residual sum of squares, which are the factors from an estimated model for model selection. Finally, we can use the adequate ARIMA model to forecast the future electricity data. After forecasting electricity data, the future electricity situation such as surplus or deficient in electricity can be estimated.

3.3 Genetic Algorithm Optimization for ESS Scheduling

In general, the deficiency in electricity in a MG can be satisfied by purchasing electricity from the utility and/or other MGs and/or by discharging the ESS. However, the charging/discharging of ESS will shorten the lifetime of ESS, as a result of which additional cost is incurred on the batteries for purchase of new ESS. Thus, in this work, the charging/discharging of ESS is associated with an additional cost penalty. Our goal is to reduce the overall electricity cost, while considering the ESS lifetime at the same time. In this subsection, we will illustrate the formulation of optimization problem. Then, the process of GA optimization will be explained.

Proposed Cost Function. To find a feasible schedule for a set of micro-grids such that an appropriate trade-off is made between the total electricity cost and the ESS lifetime, the target optimization problem is formulated as follows, where the cost function is specified under three constraints for each micro-grid.

$$\max \sum_{n=1}^{N} \Big(\alpha_{save} \big[(BPrice_{MG,n} \times GenSup_{MG,n}) + (BPrice_{U,n} \times GenSup_{U,n}) - (SPrice_{MG,n} \times MGSup_{L,n}) - SPrice_{U,n} \times (USup_{L,n} + USup_{ESS,n}) \big] - \alpha_{life} \big[Cost_{ESS} \times LOL_n \big] \Big) \tag{4}$$

subject to

$$1 \leq n \leq N$$
$$CSOC_{min} \leq CSOC_n \leq CSOC_{max}$$
$$Gen_{total,n} = GenSup_{L,n} + GenSup_{ESS,n} + GenSup_{MG,n} + GenSup_{U,n}$$
$$Load_{total,n} = GenSup_{L,n} + ESSSup_{L,n} + MGSup_{L,n} + USup_{L,n}$$

For next time slot ($n = 1$), a schedule is generated by considering future N time slots, where $N > 0$. The parameter α_{life} is a weight factor, and is set to ten times of $Capacity_{Battery}$, where $Capacity_{Battery}$ is the energy capacity value of each battery in ESS. α_{save} is a constant, and is set to 1. The second constraint is to ensure that the $CSOC_n$ of ESS is between the lower bound ($CSOC_{min}$) and the upper bound ($CSOC_{max}$) at each time slot n. The third and fourth constraints are to ensure that the electricity power is balanced between the generation and the load at each time slot n, where $Gen_{total,n}$ is the total generation, and $Load_{total,n}$ is the total demand loads.

With the above formulation, positive values of the cost function would indicate a profit on selling surplus electricity. Two parameters representing the surplus amount of electricity sold include the amount of generation (kWh) sold to other MGs, $GenSup_{MG,n}$, and that sold to the utility denoted by $GenSup_{U,n}$. Negative values of the cost function represent the cost incurred by buying deficient electricity in an MG, and/or when the ESS needs to be charged. The two parameters representing the amount of deficient electricity bought include $MGSup_{L,n}$ the capacity (kWh) supplied by other MGs to loads, and $USup_{L,n}$ is that supplied by the utility to demand loads. $USup_{ESS,n}$ is the capacity (kWh) that the utility supplies to ESS. The loss of life (LoL) of ESS [21] is affected by three factors, namely electricity usage, the amount of generated power (kWh) used to charge ESS $GenSup_{(ESS)}$, the amount of capacity (kWh) that ESS discharges for demand loads $ESSSup_L$, and $USup_{ESS,n}$.

The Process of GA Optimization. After prediction models have predicted the electricity data for the consumption of power loads and the generation of renewable energy resources, the process of optimization can be performed. First, we will randomly generate the initial set of schedules. Pairs of the schedules are selected as parents, which are merged into a single schedule during the crossover and mutation operations. A new schedule set evolves by replacing a schedule in the old set. This process is repeated until the optimization time is finished or the schedule set converges. Finally, we can get the optimal trading electricity capacity and charging/discharging capacity of ESS for each micro-grid.

First, randomly generate the initial set of schedules. As shown in Fig. 2, a schedule is encoded by a three dimensional matrix, where rows, columns, and pages represent micro-grids, units need to be scheduled, and N time periods in the future, respectively. Each micro-grid has its own schedulable units. For example, in Fig. 2, the schedulable units are $GenSup_{L,n}$, $GenSup_{ESS,n}$, $GenSup_{MG,n}$, $GenSup_{U,n}$, $ESSSup_{L,n}$, $MGSup_{L,n}$, $USup_{L,n}$, and $USup_{ESS,n}$. The values in the matrix represent electricity power (kWh), and generated randomly, but satisfying the constraints given in the cost function (Eq. 4).

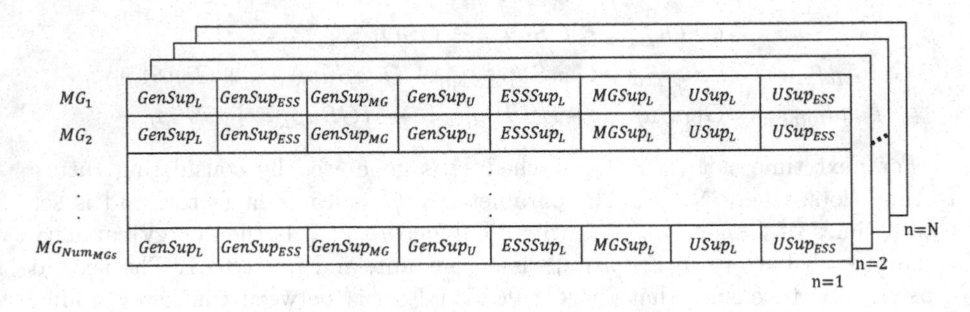

Fig. 2. The initial set of a schedule

Second, a selection operation will choose two schedules in the current scheduling set. The chosen schedules will be the parents of the next generation. This work adopts the most common way, which gives the best schedule a high probability to be selected than the worst one. The value of each schedule is calculated by the cost function (Eq. 4). In other words, a schedule with the highest cost will be the best schedule in the scheduling set. After evaluating a schedule, its normalized rank is calculated [12]. The normalized rank is used to adjust the probability of a schedule being selected. It is enforced that the probability of the best schedule to be selected is four times more than that of the worst.

After parent selection, a crossover is executed. We use a classical one-point crossover operation by cutting the pages of a matrix as shown in Fig. 2 in random to generate a crossover point, and then exchanging segments of the two parents to produce a child. After crossover operation, we should adjust some schedulable units to satisfy constraints of the cost function. A schedule produced via the classical one-point crossover would be an infeasible one that relates the constraints given in Eq. 4. Thus, an adjustment is made such that a child completely inherits part of schedulable units from parent 1, and the remaining schedulable units are filled by parent 2. This adjustment plays the role of mutation.

When a new child schedule is generated, it will substitute a schedule in the current scheduling set to generate a new scheduling set. The proposed method replaces the schedule that has the lowest cost in the current scheduling set by the newly generated one.

4 Experiments

In this section, we present evaluation of the proposed MPO method for smart grids. First, the experimental setup used for experiments is introduced. Then, the experimental results are illustrated.

4.1 Experimental Setup

In this subsection, the experimental environment and energy data used in our experiments are described.

Experimental Environment. The proposed method is implemented in the Python and Matlab programming language on a PC with Intel(R) Core(TM) i7 2.39 GHz CPU, 4 GB RAM running Windows 7 64-bit OS. The Python programming language is used for implementing the weather information parser such as irradiance and wind speed. The Matlab programming language is used for realizing the proposed distribution management.

Demand Load Data, Generation Data, Electricity Price, and Power Proportion. Three different types power consumers including commercial consumers, industrial consumers, and residential consumers are considered. We refer the different conditions of electricity usage with different times in research [23] to simulate one day consumption for each type of demand load. The demand load percentages for each type of consumers are commercial consumers 13%, industrial consumers 73%, and residential consumers 14%. By observing the usage frequency of demand loads throughout a day, demand loads behavior can be classified into peak time and non-peak time. To reduce peak load demands, the utility sets the price higher at peak time than at non-peak time. In this work, the utility selling price is set to \$3.62 per kWh at peak time and \$1.69 at non-peak time. However, the utility buying price is set to \$2 per kWh at peak time and \$1 at non-peak time. Between all micro-grids, the trading price of selling and buying are the same. They are set to \$3 per kWh at peak time, and \$1.5 per kWh at non-peak time.

4.2 Experimental Results

In this subsection, we first give an evaluation of the ARIMA prediction model. Then, the cost saving of electricity and ESS LoL for smart grids due to the proposed method are described.

Evaluation of ARIMA Prediction Model. We use the ARIMA model to predict the demand loads, wind speed, and irradiance. For predicting the values precisely, the past 100 hours history data are used. The prediction accuracy is evaluated by the Root Mean Squared Error (RMSE) as shown in Eq. 5. RMSE makes an excellent general purpose error metric for a prediction method.

$$RMSE = \sqrt{\frac{1}{n}\sum_{i=1}^{n}(y_i - \widehat{y_i})^2} \tag{5}$$

where n is the number of predicted data, y_i is the real data, and $\widehat{y_i}$ is the predicted data. The smaller the RMSE value, the higher the prediction accuracy is. To normalize the RMSE, Eq. 6 is used to explain the error rate between different scales of data, where the maximum error is the difference between the maximum and minimum value of data. The error rates of demand loads, wind speed, and irradiance are 4.77%, 1.60%, and 7.18%, respectively. We can observe that all error rates are smaller than 10%, which is quite good.

$$ErrorRate = (RMSE/MaximumError) \times 100\% \tag{6}$$

Optimization Method. In the following experiments, we assume the smart grid has 30 micro-grids. First, we consider several look-ahead time slots, such as 2, 3, 4, 5, and 6 time slots and conclude on an appropriate number of time slots to use for MPO. Finally, electricity cost, ESS LoL, and overall cost are compared to the MPC look ahead-dispatch [14] which is also MPC based method for electricity cost saving, and economic dispatch system [13] which is state-of-the-art method for electricity cost saving.

First, we consider several look-ahead time slots and conclude on an appropriate number of time slots to use for MPO. As shown in Fig. 3, we analyze different look-ahead time slots to be used for the proposed MPO method. The electricity

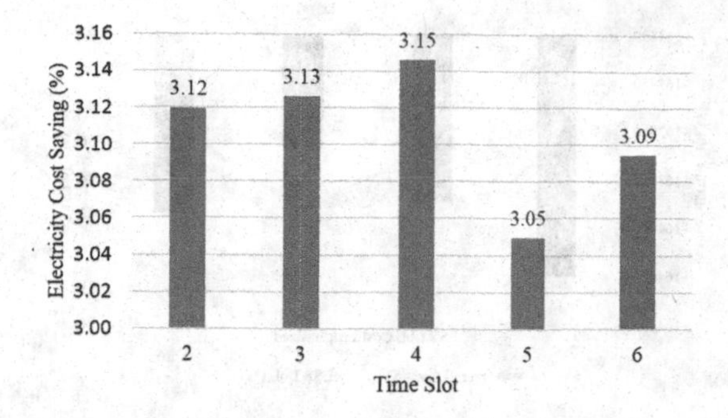

Fig. 3. Cost saving in different look-ahead time slots

cost saved by MPO is compared to that by traditional method. In the traditional method, the ESS is used as an uninterruptible power supply (UPS), that is, for backup purposes only, and trading only considers the electricity demand response for the current time slot. From Fig. 3, we can observe that the smart grid saves 3.15% cost if four time slots are used in MPO, which is the highest saving. Thus, 4 time slots should be an appropriate size of prediction horizon for MPO.

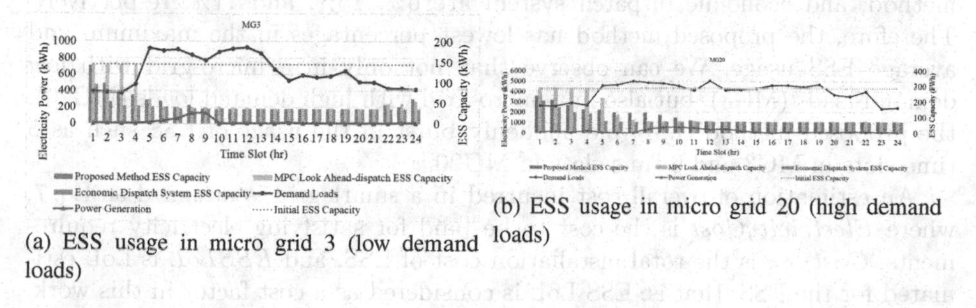

(a) ESS usage in micro grid 3 (low demand loads)

(b) ESS usage in micro grid 20 (high demand loads)

Fig. 4. Comparison of ESS usage with other methods

Finally, electricity cost, ESS LoL, and overall cost are compared to other methods. Figure 4 shows the comparison of ESS usage in one day, for MPO, MPC look ahead-dispatch method, and economic dispatch system. In the MPC look ahead-dispatch method [14], ESS not only plays a backup role, but also acts as an active electricity supplier. ESS will be discharged as demand loads are higher than a given threshold, where the threshold is set to the average of past demand loads, and will be charged as demand loads are lower or equal to the given threshold. In the economic dispatch system [13], ESS will be discharged/charged to minimize the cost of electricity. Both of them also use predicted future electricity to determine the trading electricity amount.

Figure 4 shows the demand loads (kWh) and power generation (kWh) varying with the hour of the day along with the ESS capacity for two different

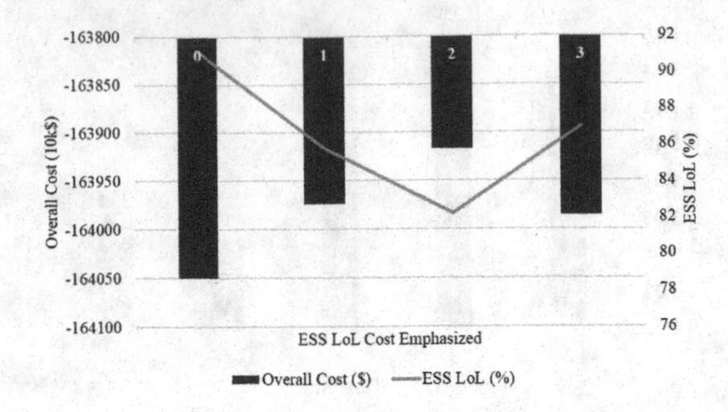

Fig. 5. ESS evaluation with different weight factors for α_{life}

micro-grids, one with a low demand load consumption (MG3) and one with a high demand load consumption (MG20). In MG3, the maximum ESS usage percentages of MPO, MPC look ahead-dispatch method, and economic dispatch system are 30%, 59%, and 35%. The average ESS usage percentages of MPO, MPC look ahead-dispatch method, and economic dispatch system are 6%, 10%, and 12%, respectively. In MG20, the maximum ESS usage percentages of MPO, MPC look ahead-dispatch method, and economic dispatch system are 12%, 37%, and 28%. The average ESS usage percentages of MPO, MPC look ahead-dispatch method, and economic dispatch system are 6%, 15%, and 11%, respectively. Therefore, the proposed method has lowest percentages in the maximum and average ESS usage. We can observe that not only in a micro-grid with low demand loads (MG3), but also in a micro-grid with high demand loads (MG20), the proposed method can avoid suddenly burst in the usage of ESS such as 5 time slots in MG3 and 6 time slots in MG20.

An estimation of overall cost incurred in a smart grid is designed as Eq. 7, where *ElectricityCost* is the cost to be paid for satisfying electricity requirements, $Cost_{ESS}$ is the total installation cost of ESS, and *ESSLoL* is LoL estimated for the ESS, that is, ESS LoL is considered as a cost factor in this work.

$$OverallCost = ElectricityCost + (Cost_{ESS} * ESSLoL) \quad (7)$$

To evaluate the impact of overall cost and ESS LoL with different weight factors α_{life}, the simulation results are shown in Fig. 5. We can observe that giving two times emphasis to α_{life} results in the least overall cost and ESS LoL (i.e. 82%). However, giving three times emphasis to α_{life}, leads to a higher overall cost and ESS LoL (i.e. 87%). Because of discharging/charging ESS is too expensive, ESS will not be discharged/charged at non-peak time, and be discharged for a higher capacity at peak-time with a limited discharged capacity. Therefore, the more emphasis consumers put on ESS, it will not always lead to a better overall cost and ESS LoL. Compared to the MPC look ahead-dispatch method, the overall cost saving and ESS LoL reduction achieved by MPO is 0.85% (i.e. $1,362,000) and 12.18% for 20 months, respectively.

The long term trend of ESS LoL and overall cost saving are shown in Fig. 6. To compare the overall cost with other methods, we set the overall cost of

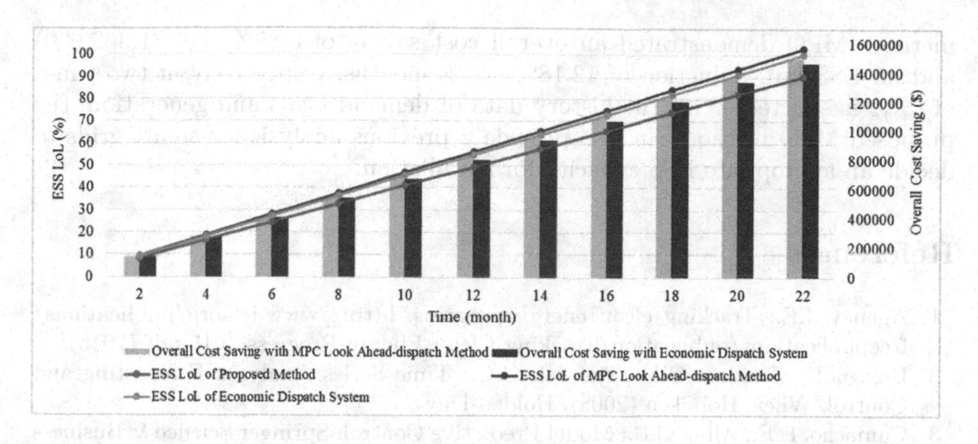

Fig. 6. Comparison of ESS LoL and overall cost saving with state-of-the-art methods

proposed MPO method as a baseline, and observe the difference with other methods which is called as the overall cost saving ($). In Fig. 6, we can observe that the overall cost savings ($) of proposed MPO method compared with other methods grow gradually with time. After 22 months, ESS LoL in MPO, the MPC Look Ahead-dispatch Method, and the Economic Dispatch System are 94%, 103%, and 100%, respectively. In other words, ESS in the proposed MPO method has a longer life than that in other methods. Even other methods have run out of ESS lifetime, the proposed MPO method still has 6% ESS lifetime to be used. Moreover, to compare the overall cost with MPC look ahead-dispatch method and economic dispatch system, the proposed method can save $1,498,200 and $1,467,840, respectively. Therefore, to save the overall cost and extend ESS lifetime, it is better to consider ESS LoL while using it as an active supply.

5 Conclusions

In this work, we proposed a model predictive optimization (MPO) method for distributed management system in smart grids. MPO leverages prediction models for demand loads and power generation, while scheduling the charging/discharging of energy storage systems (ESS) and the trading of electricity among micro-grids with each other and with the utility, using ESS as an active power supply. The Autoregressive Integrated Moving Average (ARIMA) model was used to predict the energy states of power loads and renewable energy resources. Our prediction model exhibited a RMSE error of less than 10% for predicting demand loads and renewable energy generation. The load and generation prediction results were used as inputs to the optimizer. The Genetic Algorithm (GA) optimizer was used to schedule the usage of ESS, and to reduce the overall cost, which includes the total electricity cost and the cost incurred due to ESS loss of life (LoL). Experimental results showed that 100 schedules in the initial scheduling set and 4 look-ahead time slots are appropriate settings for the optimizer. Compared with the model predictive control (MPC) look ahead-dispatch method, MPO demonstrated an overall cost saving of 0.85% (i.e. $1,362,000) and an ESS LoL reduction of 12.18% for 20 months, respectively, if two times of α_{life} is selected. With the history data of demand loads and generation, the proposed MPO method can also provide a previous analysis for smart grids to decide an appropriate ESS capacity for installation.

References

1. Agency, I.E.: Tracking clean energy progress. http://www.iea.org/publications/freepublications/publication/Tracking_Clean_Energy_Progress_2015.pdf (2016)
2. Box, G.E., Jenkins, G.M., Reinsel, G.C.: Time Series Analysis: Forecasting and Control. Wiley, Hoboken (2008). Holden-Day
3. Camacho, E.F., Alba, C.B.: Model Predictive Control. Springer Science & Business Media, London (2013). https://doi.org/10.1007/978-0-85729-398-5
4. Chung, S.H., Chan, H.K.: A two-level genetic algorithm to determine production frequencies for economic lot scheduling problem. IEEE Trans. Indus. Electron. **59**(1), 611–619 (2012)
5. Department of Energy: The Smart Grid: An Introduction (2016). http://energy.gov/oe/downloads/smart-grid-introduction-0
6. University of Florida: Systems and Optimization Aspects of Smart Grid Challenges (2016). http://www.ise.ufl.edu/SGC2011/SG_Conference/Welcome.html
7. Glover, F., Laguna, M.: Tabu search. In: Du, D.Z., Pardalos, P.M. (eds.) Handbook of Combinatorial Optimization. Springer, Boston (1998). https://doi.org/10.1007/978-1-4613-0303-9_33
8. Holkar, K., Waghmare, L.: An overview of model predictive control. Int. J. Control Autom. **3**(4), 47–63 (2010)
9. Hu, S.: Akaike information criterion. Center for Research in Scientific Computation (2007)
10. Kuck, M., Scholz-Reiter, B.: A genetic algorithm to optimize lazy learning parameters for the prediction of customer demands. In: Proceedings of the IEEE Machine Learning and Applications (ICMLA), pp. 160–165 (2013)
11. Lawson, B.: Battery life (and death). http://mpoweruk.com/life.htm (2016)
12. Lee, E., Bahn, H.: A genetic algorithm based power consumption scheduling in smart grid buildings. In: Proceedings of the International Conference on Information Networking (ICOIN), pp. 469–474 (2014)
13. Mahmoodi, M., Shamsi, P., Fahimi, B.: Economic dispatch of a hybrid microgrid with distributed energy storage. IEEE Trans. Smart Grid **6**(6), 2607–2614 (2015)
14. Mayhorn, E., Kalsi, K., Elizondo, M., Zhang, W., Lu, S., Samaan, N., Butler-Purry, K.: Optimal control of distributed energy resources using model predictive control. In: Proceedings of the IEEE Power and Energy Society General Meeting, pp. 1–8 (2012)
15. Miao, H., Huang, X., Chen, G.: A genetic evolutionary task scheduling method for energy efficiency in smart homes. Int. Rev. Electr. Eng. **7**(5), 5897–5904 (2012)
16. Otomega, B., Marinakis, A., Glavic, M., Cutsem, T.V.: Model predictive control to alleviate thermal overloads. IEEE Trans. Power Syst. **22**(3), 1384–1385 (2007)
17. Parisio, A., Rikos, E., Glielmo, L.: A model predictive control approach to microgrid operation optimization. IEEE Trans. Control Syst. Technol. **22**(5), 1813–1827 (2014)

18. Peek, G.H., Hanley, C., Boyes, J.: Solar energy grid integration systems - energy storage. Sandia Report, SAND2008-4247 (2008)
19. Qin, S.J., Badgwell, T.A.: An overview of industrial model predictive control technology. In: Proceedings of the AIChE Symposium Series, vol. 93, pp. 232–256 (1997)
20. Schwarz, G.: Estimating the dimension of a model. Ann. Statist. **6**(2), 461–464 (1978)
21. Tran, D., Khambadkone, A.M.: Energy management for lifetime extension of energy storage system in micro-grid applications. IEEE Trans. Smart Grid **4**(3), 1289–1296 (2013)
22. Van, L., Peter, J., Aarts, E.H.: Simulated Annealing. Springer, Heidelberg (1987). https://doi.org/10.1007/978-94-015-7744-1
23. Vattekar, E.: Analysis and model of consumption patterns and solar energy potentials for residential area smart grid cells. Master's thesis, Norwegian University (2014)
24. Yildirim, M.B., Mouzon, G.: Single-machine sustainable production planning to minimize total energy consumption and total completion time using a multiple objective genetic algorithm. IEEE Trans. Eng. Manag. **59**(4), 585–597 (2012)

Attacker Evidence System in WSN

B. Srinivasa Rao[1(✉)] and P. Premchand[2]

[1] Department of Computer Science and Engineering,
Gokaraju Rangaraju Institute of Engineering and Technology (Autonomous)
(Affiliated to Jawaharlal Nehru Technological University Hyderabad,
Hyderabad 500072, India), Bachupally, Hyderabad 50090, Telangana, India
bsrgriet2015@gmail.com

[2] Department of Computer Science Engineering,
University College of Engineering, Osmania University,
Hyderabad 500007, India
p.premchand@uceou.edu

Abstract. Due to the features like distributed structure, open wireless network system etc. the Wireless Sensor Networks (WSN) are pruned to security attacks at various levels. These attacks may have significant influence on the efficiency of WSN. During the anomalous attacks, attackers manage to get unauthorized accesses to the network and harm the network system and services to make them ineffective. A counter mechanism is essential to overcome the influence of the attacks and sustain the efficiency of the network. In that process it is required to find the evidence for the activities of the attacker in the network. In the present research work, an attempt has been made to develop and implement a mechanism or scheme to find the evidence for the existence of an attacker in the network and to provide security measure to the WSN system by filtering the attacker to prevent the attacks. This is achieved by designing and implementing an Attacker Evidence System (AES) as a simple network security measure in wireless sensor networks systems. The proposed AES is designed for homogeneous and heterogeneous WSN models considering single and multiple-sensing detection schemes. The present security measure and its simulation results have been presented and discussed. The results reveal that the present AES works as per expectations for both the types WSNs and can be a proto-type for further extensions.

Keywords: Information security · Network security · Attack · Attacker
Intrusion detection · Intruders · WSN · Heterogeneous

1 Introduction

Now-a-days, it has become essential for every organization to have its own security policy as per its requirements based upon its adopted technology like Communication Network, Parallel Computing System, Distributed Computing System, Cloud System, Adhoc Network, Mobile Network, Wireless Sensor Network etc. This security policy may be intended to protect organization through pro-active policy stance [1]. From the literature it is well understood that Computer Security is concerned with the loss or

R. Sharma et al. (Eds.): ICAN 2017, CCIS 805, pp. 166–178, 2018.
https://doi.org/10.1007/978-981-13-0755-3_13

harm to the hardware, software or information of an organization. It also includes denial, disruption and misdirection of the services and facilities provided by the computer system [2–6]. The Computer Security may be considered as combination of System Security, Network Security and Data or Information Security. Data security or Information Security deals with security issues, policies and services of data under communication. Data Security provides security services for threats concerned with data confidentiality, authentication, integrity, non-repudiation, access control and availability [7–10]. As Information Systems are designed in multilayered structures, the above security issues have their influence at different layers of the systems and affect the performance of the Systems [11]. In this context, the security issue like confidentiality is becoming a challenge task in the environment of new technologies such as cloud computing, wireless communication systems etc. [12]. One aspect of the confidentiality of an Information System is unauthorized access to the network by a third party to steal important information or causing damage to the efficiency of the Information System [7–10]. An unauthorized access to the computer networking system is known as attack/hack/intrusion and is one of the most serious threats to the Computer Security. Hence, it is essential to design a security measure to detect the attacker to assess the vulnerability of the system or to protect the system from misuse [7]. An Attacker Evidence system (AES) is software and/or hardware based security scheme to detect the attempts of an attacker intended to misuse the systems such as network or the Internet [13].

A wireless sensor networks (WSN) is a wireless network consisting of spatially distributed autonomous devices using sensors to cooperatively monitor physical or environmental conditions [14]. The WSN have many applications such as military, civil, healthcare, home automation, traffic control etc. It normally constitutes a wireless adhoc network associating with a multi-hop routing algorithm [15]. A WSN is an adhoc distributed system consisting of several wirelessly connected sensor nodes and can be deployed to collect information about surrounding environment [16]. WSNs are highly vulnerable to security attacks at various levels due to various factors like distributed nature, multi-hops, open wireless medium etc. [16–18]. Hence an effective security measure is to be designed to overcome the attacks like intrusion or hacking in WSN. An Attacker Evidence system (AES) can be designed and implemented to detect and prevent from security attacks [19]. Survey reveals that earlier, several researchers have designed and implemented Intrusion Detection Systems for WSN in different scenarios such as Anomaly-based IDS, Signature-based IDS, and Cross layer IDS etc. [13, 17, 18]. The probability of creating more false alarms is a problem with Anomaly-based IDSs, even though they are lightweight in nature. Overheads like updating and inserting new signatures and suitability to larger WSN are the disadvantages with Signature-based IDSs. As the WSNs have resource limitation, the Cross layer IDSs are usually not suitable [13, 17, 18]. Based upon the capability the WSNs can be classified as homogeneous and heterogeneous. Large sensing range, more power and broad casting power management information are the significant features of Heterogeneous WSNs in comparison with homogeneous WSNs [13, 14]. The two important conditions for ensuring detection probability in WSNs are the network connectivity and broad cast reach ability in a secured manner [14, 18, 20]. A few have considered the case of IDS for heterogeneous WSN security in comparison with homogeneous one with a simple simulation method. A comparative study may be considered for both homogeneous and heterogeneous WSNs in terms of intrusion/hacker detection. Hence this is the motivation for the present work to design and implement an Attacker Evidence system (AES) for homogeneous and heterogeneous WSNs by using a simple simulation method. This simple method may be a proto-type but would be useful to extend further. To the best of our knowledge, our effort is the first to address this issue both in homogeneous and heterogeneous WSNs for a simple simulation using Attacker Evidence system (AES).

2 Earlier Intrusion Detection Systems (IDS) and WSN

Various attacker/intrusion/hacker detection systems have been designed and implemented in different scenario and detailed information is available in vast literature [13–29]. It is already understood that An Intrusion Detection system (IDS) is software and/or hardware based security scheme to detect the attempts of an intruder intended to misuse the systems such as network or the Internet [13]. From [14], the IDS comprise of mainly three components namely sensors, console and central engine. The security events of the WSN are produced by sensors. The WSN events and their related alerts are monitored by console. The centrals records events and set rules for generation of alerts. The intrusion detection is possible in two ways: intrusion detection by a single sensor or multiple sensors collective cooperation. As the former is ineffective in some cases, multiple sensor detection can be considered for intrusion detection. The data flow in homogeneous and heterogeneous wireless sensors is as shown in Fig. 1. S and D indicate Source and Detector and R1, R2, and R3 are receiving nodes in WSN. The directions indicate the flow of data through the networks. The intruder may be denoted by a cloud symbol.

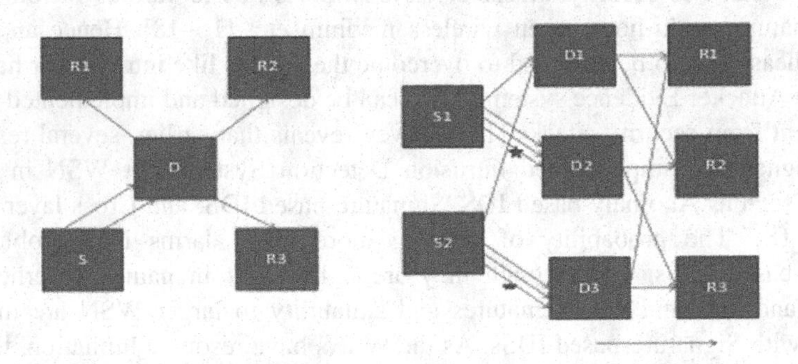

Fig. 1. Homogeneous and Heterogeneous WSNs

3 Attacker Evidence System (AES)

With reference to [14], the presently proposed simple Attacker Evidence System (AES) can be designed in five modules: 1. WSN construction, 2. Generation of Packets. 3. Identifying authorized and unauthorized port. 4. Inter-Domain Packet Filter construction and 5. Valid packet reception. In the first module WSN is designed such a way that each node is connected to the neighboring nodes and each port number is authorized by all nodes. In the second module a browser is designed to convert selected data into a fixed size of packet. These packets are sent from source to detector. In the third module in order to find authorized and unauthorized port a detection mechanism is designed. This module checks whether the path is authorized or unauthorized using the port number and if path is authorized the packet is send to valid destination. Otherwise the packet will be deleted. In the fourth module the Inter-Domain Packet Filter is designed. The Packet Filter filters the packets received from other than the designated port number and authorized packets will be send to destination. Finally, the valid packet reception module receives all the valid Packets. Thus only valid packets reach the destination from the source node [30]. The design logic for the Attacker Evidence System (AES) is shown in Fig. 2. The system design comprises of mainly data input and output mechanisms. 1. Input Design: (i) Source file browsing (ii) Conversion of selected data into fixed size packets. (iii) Write program to hack the packet (iv) Selection of port number to send the packet (v) Sending packet from source to detector. 2. Output Design: (i) Filtering and discarding of packet from unauthorized port (ii) Sending authorized packets to destination. The functional flow of data, data input, intruder detection, packet filtering, and reception packets are shown in data flow diagram (Fig. 2).

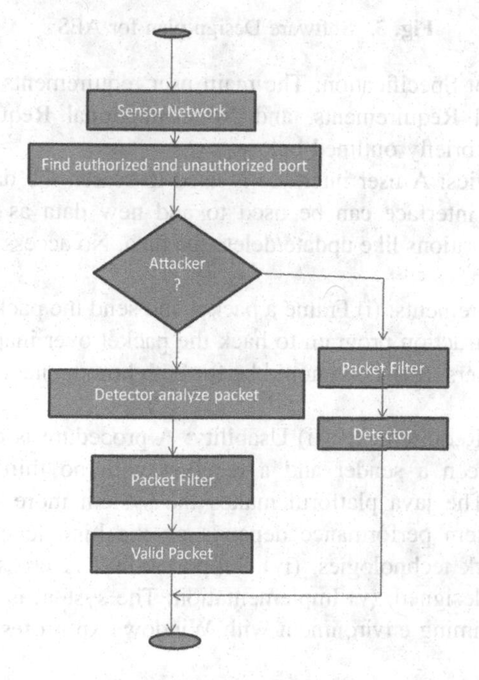

Fig. 2. Data flow diagram for Attacker Evidence System (AES)

At first the user will input the data from a file and sends this packet to the detector and the detector filters the received packets. In case the packet is authorized it will be sent to a valid receiver. If the packet is an unauthorized one, then it will be discarded into the sink. Thus the design plan is implemented in four modules: Network construction module, Detector module, Packet filter module and Receive packet module. The corresponding software design plan is shown in Figs. 3 and 4. The Network Construct module is a network, with attributes Construct and with responsibilities container.add(c); The Detector module comprises the attributes analyzing and responsibilities void server(); The packet filter contains attributes Testing and responsibilites r1.server; Finally the sink module contains attributes Receive packets and responsibilites get.packet().

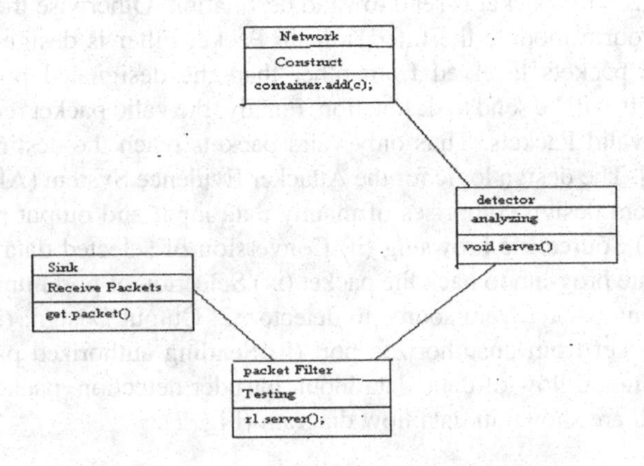

Fig. 3. Software Design plan for AES

(a) User Requirement Specification: The main user requirements are User Characteristics, Functional Requirements, and Non Functional Requirements. The user requirements are briefly outlined below:

1. User Characteristics: A user interface is used to search the data and services. An operational user interface can be used to add new data as and when required. Provision for operations like update/delete the data. No access rights for the user to access the system.
2. Functional Requirements: (i) Frame a packet and send the packet over the network. (ii) Write the instruction program to hack the packet over inappropriate, incorrect, anomalous attackers. (iii) This should be for both homogeneous and heterogeneous WSN models.
3. Non-Functional Requirements: (i) Usability: A procedure is designed to establish connection between a sender and a receiver with no third party intervention. (ii) Reliability: The java platform makes the system more reliable. (iii) Performance: The system performance depends on the high level languages and the advanced network technologies. (iv) Supportability: A cross platform supported system is to be designed. (v) Implementation: The system is implemented in java network programming environment with Windows xp professional platform.

(b) System Requirements: The system requirements for the AES are: Hardware: Pentium IV 2.6 GHz Processor, 512 MB DD RAM, 20 GB Hard Disk, LG 52X CD Drive, Standard Keyboard, Mouse, Internet/Networking. Software: Java, JFrameBuilder and Window's Xp.

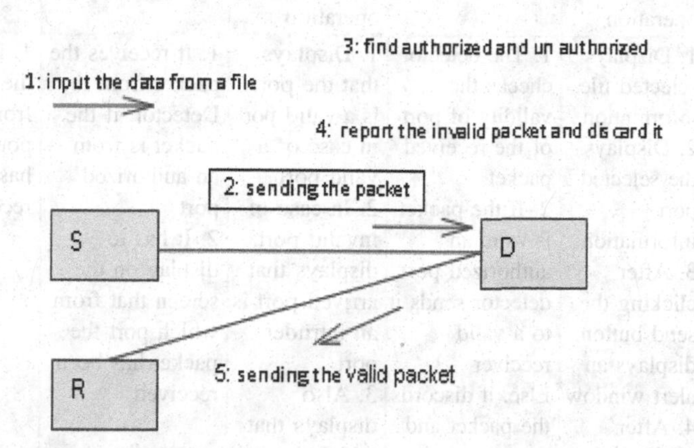

Fig. 4. Mechanism for receiving valid packets

4 AES Implementation

The architecture of a WSN node is as shown in Fig. 5. According to the networking principles each node contains the data of authorized ports of all other nodes in the network. Each node can verify whether a packet is from an authorized port or not by running a suitable algorithm and accordingly takes the decision for next action. All the operations, respective screen display operations and screen displays involved in the simulation at source, detector and receiver level are presented in Table 1. Predefined authorized and unauthorized ports data has been stored in files and the files have been browsed to select the ports for communicating packets through the WSN. The present AES has been simulated in the environment of Java, JFrame Builder and Window's XP operation system using the specified hardware and software. The simulated results that have been obtained by implementing the operations as per the Table 1 have been reported in Table 2. Also important screen shots have been presented for better understanding of the simulated results and the process of AES in Fig. 6.

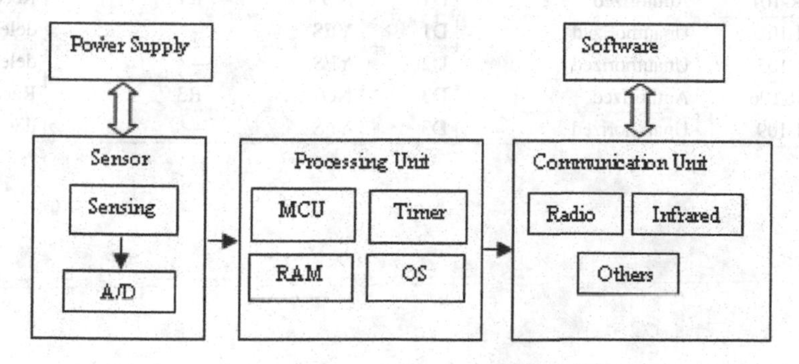

Fig. 5. Architecture of a WSN node

Table 1. Operations and Result displays on Computer during the simulations

Source		Detector		Receiver	
Operations	Screen display operation	Operations	Screen displays operation	Operations	Screen displays
1. Double click on source batch file 2. Click on browse button to select the file 3. Select the port number 4. Click the send button 5. Click on the OK button	1. Displays selected file information 2. Displays the selected port information 3. After clicking the send button, displays an alert window 4. After clicking the OK button the packet is sent to the Detector	1. The detector checks the validity of port of the received packet 2. If the packet is from an authorized port, detector sends it to a valid receiver Else, it discords the packet and reports the intrusion	1. Displays that the port is a valid port in case of a valid port 2. In case of invalid port, displays that arrived port is an intruder port 3. Also displays that which detector has detected	1. It receives the packet from the Detector, if the packet is from an authorized port 2. It has to display on the screen that from which port the packet has been received	1. Displays on the screen that from which port the packet has been received

Table 2. Simulated results of attacker evidence scheme for both Homogeneous and Heterogeneous WSN

Source	Selected port	Port Authorized/Unauthorized	Detector	Hacking detection	Packet receiver/sink	Packet status
Homogeneous network						
S	R-101	Authorized	D	NO	R1	Received
S	I-104	Unauthorized	D	YES	–	deleted
Heterogeneous network						
S1	R-101	Authorized	D1	NO	R1	Received
S1	I-102	Unauthorized	D1	YES	–	deleted
S1	I-105	Unauthorized	D2	YES	–	deleted
S2	R-106	Authorized	D3	NO	R3	Received
S2	I-109	Unauthorized	D3	YES	–	deleted

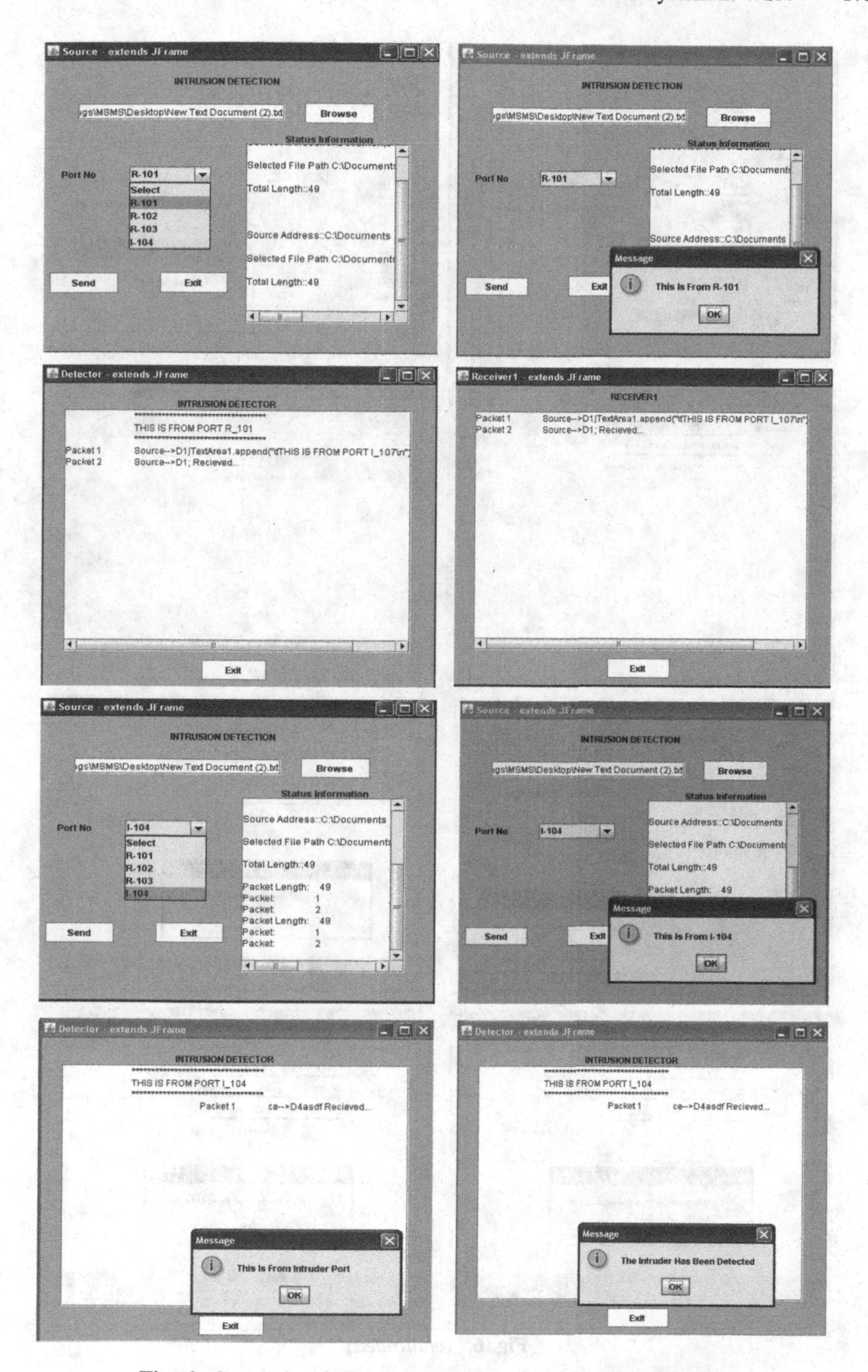

Fig. 6. Screenshot for Homogeneous and Heterogeneous WSN

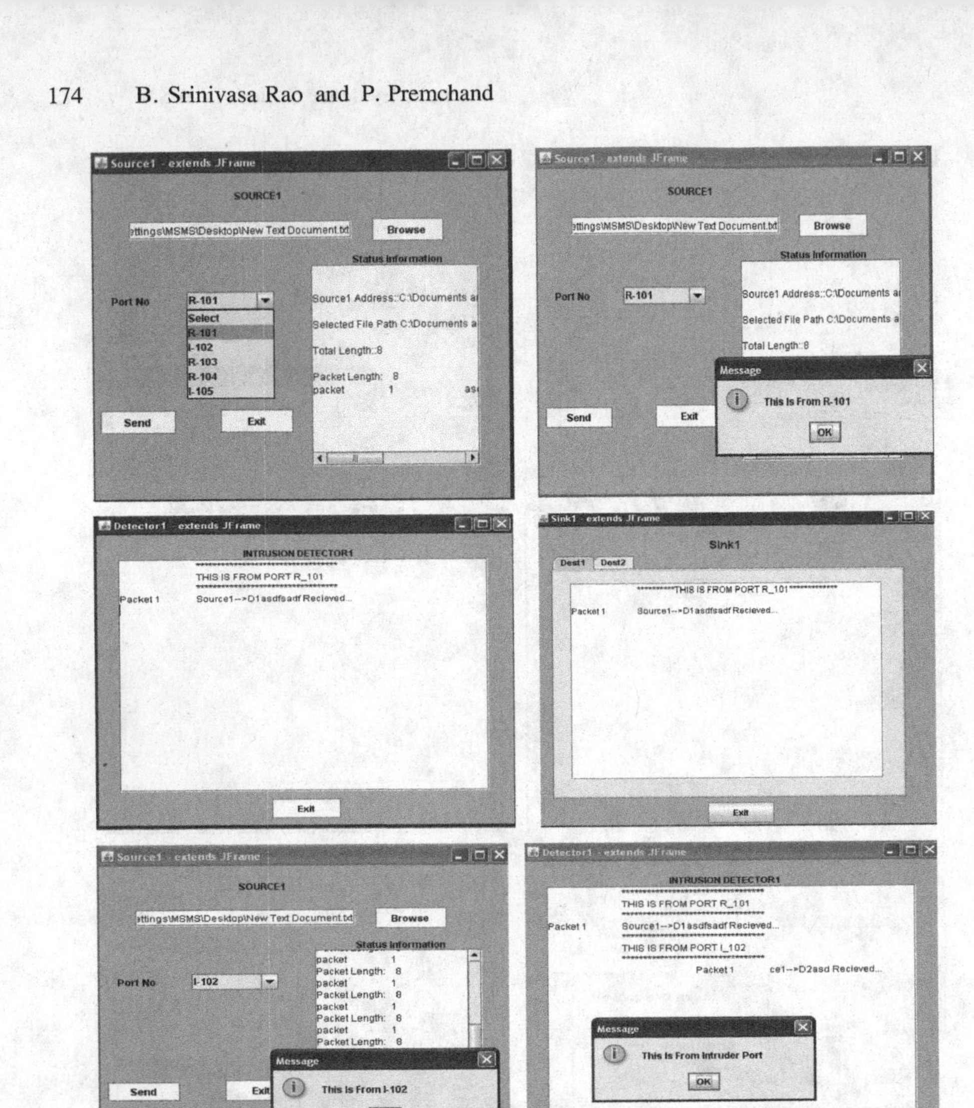

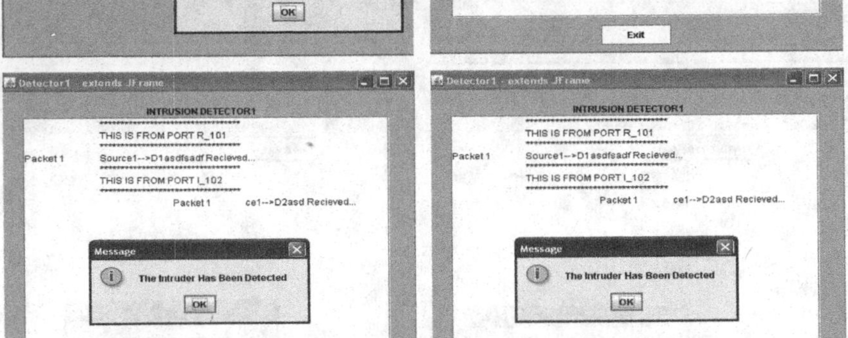

Fig. 6. *(continued)*

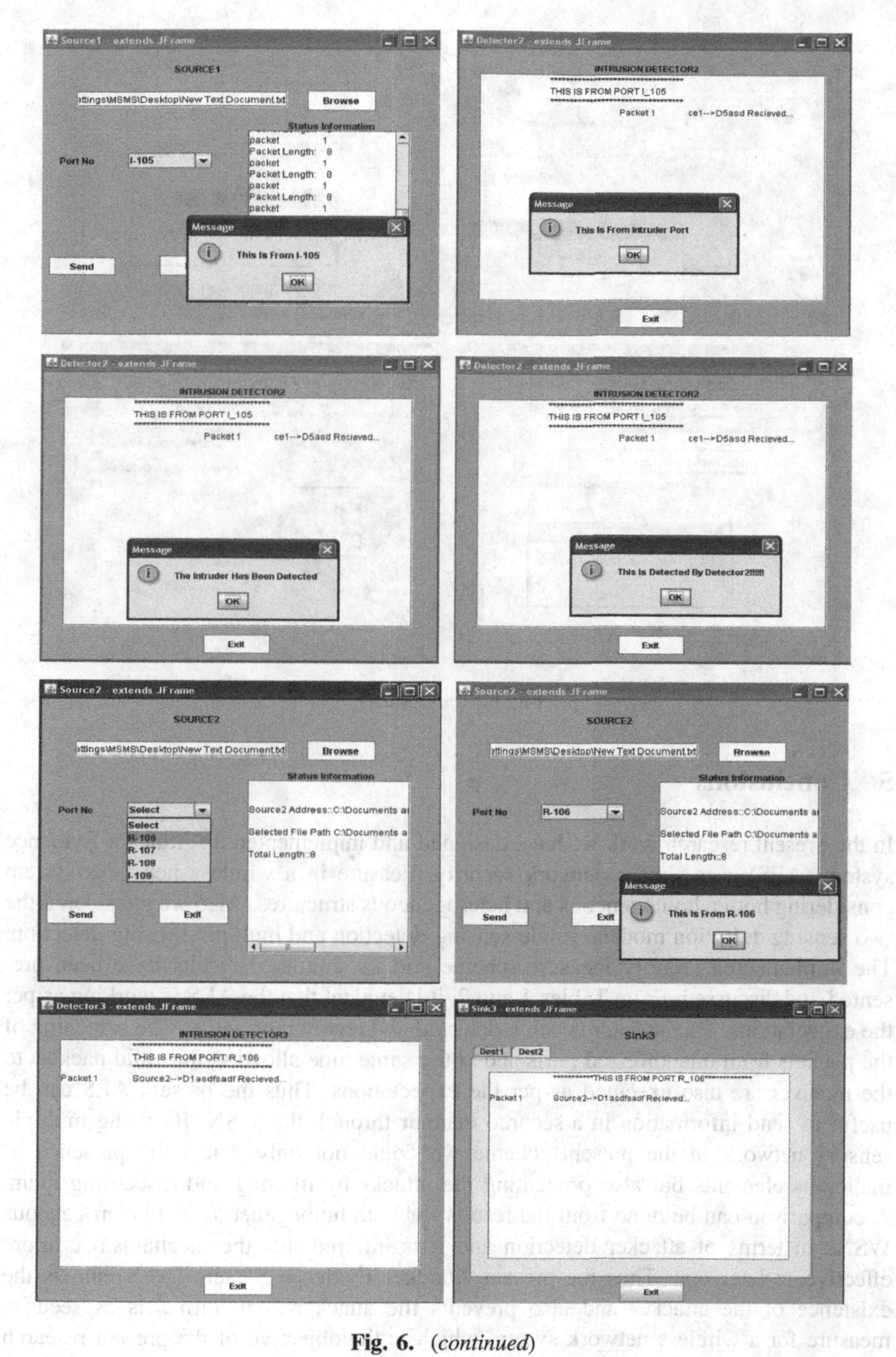

Fig. 6. (*continued*)

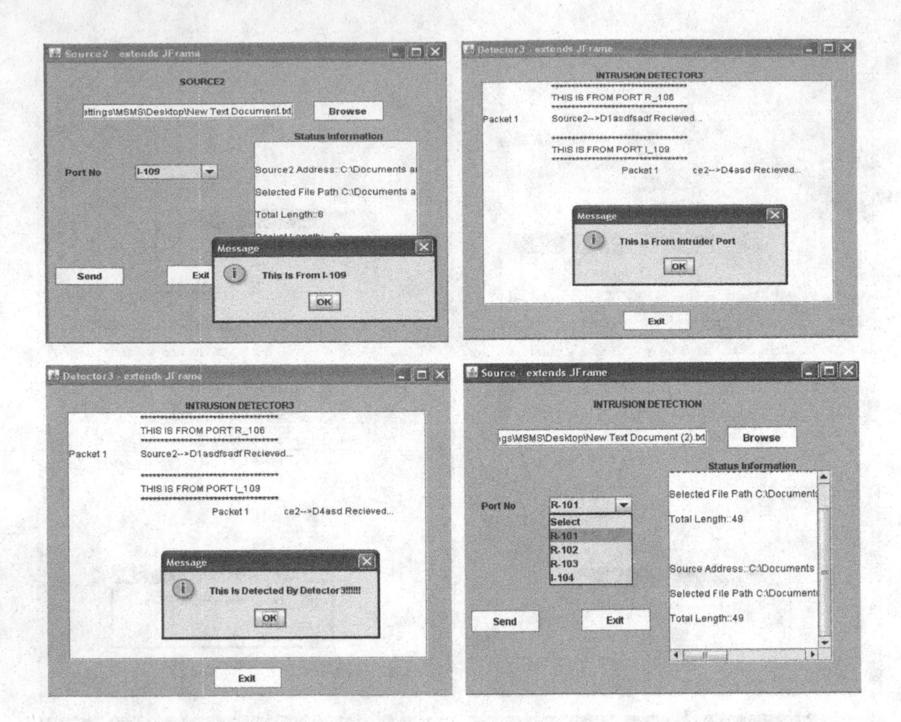

Fig. 6. (*continued*)

5 Conclusions

In the present research work we have designed and implemented an Attacker Evidence system (AES) as a simple network security measure in a wireless networks system considering both a homogeneous and heterogeneous structures. Also we considered the two sensing detection models: single-sensing detection and multiple-sensing detection. The implemented security measure scheme and its simulated results have been presented and discussed. From Tables 1 and 2, it is evident that the AES is working as per the expectations. The attacker is being detected and reported properly. The screening of the packets from unauthorized ports and at the same time allowing the valid packets to the receiver are also executed as per the expectations. Thus the present AES can be useful to send information in a secured manner through the WSN. By using multiple sensors network in the present scheme we could not only detect the presence of malicious elements but also preventing the attacks by filtering and discarding them. A comparison can be done from the results for both homogeneous and heterogeneous WSNs in terms of attacker detection and it is inferred that the mechanism is more effective in later one. Thus the present Attacker Evidence System (AES) shows the existence of the attacker and also prevents the attack and in turn acts as security measure for a wireless network system which is the objective of the present research work. In scope, the present Attacker Evidence System (AES) may be a proto-type, but the simulation can be extended to study intrusion detection probability within a certain intrusion distance under various application scenarios. The model can be further improved for a larger and more realistic WSN by characterizing attacker detection probability with respect to the intrusion distance and the network parameters like node density, sensing range, transmission range etc. The model can be further enhanced for attacker/hacker/intrusion detections in internet applications and parallel computer interconnection networks.

Acknowledgements. B. Srinivasa Rao is very much thankful to Dr. L. Pratap Reddy, Professor, Department of ECE, JNTUH, Hyderabad, for his valuable suggestions. Also thankful to the Management of GRIET for their encouragement and cooperation for pursuing his Ph.D. work.

References

1. Garret, C.: Importance of Security Policy (2012). http://www.slideshare.net/charlesgarrett/importance-of-a-security-policy-11380022
2. https://www.uniassignment.com/essay-samples/information-technology/importance-of-information-security-in-organizations-information-technology-essay.php (2017)
3. New firewall can protect your phone from security threat. http://www.deccanchronicle.com/technology/in-other-news/060717 (2017)
4. http://searchitoperations.techtarget.com/definition/hardware-security (2017)
5. https://www.itgovernance.co.uk/shop/category/information-security 2017
6. https://en.wikipedia.org/wiki/Network_security
7. Stallings, W.: Cryptography and Network Security-Principles and Practices, 4th edn. Pearson Education (2006)
8. Stallings, W.: Data and Computer Communications, 5th edn. PHI (1999)
9. Forouzan, B.A. : Cryptography and Network Security, Special Indian Edition, TMH (2007)
10. Forouzan, B.A.: TCP/IP Protocol Suite, TMH (2000)
11. Kisielnicki, A., Sroka, H.: Systemy informacyjne biznesu, Warszawa: Placet, S. 17 (2005). ISBN 83-85428-94-1
12. Wawak, S.: The importance of information security management in crisis prevention in the company. In: Proceedings of ISBAGECC-2017 (2017). http://www.academia.edu/1649676
13. Alrajeh, N.A., Khan, S., Shams, B.: A review -intrusion detection systems in wireless sensor networks. Int. J. Distrib. Sens. Netw. **2013**, 1–9 (2013). https://doi.org/10.1155/2013/304628
14. www.vidhatha.com
15. research.ijcaonline.org
16. connection.ebscohost.com
17. Agrawal, D.P., Zeng, Q.A.: Intrusion detection in wireless ad-hoc networks. In: Introduction to Wireless and Mobile Systems, 4th edn., p. 28 (2014)
18. Sharma, U., Bahl, N.: A review on security issues and attacks in wireless sensor networks. Int. J. Adv. Res. Comput. Sci. (IJARCS) **8**(4), 387–391 (2017)
19. www.hindavi.com
20. Zheng, J., Jamalipour, A. (eds.): Wireless Sensor Networks: A Networking Perspective. John Wiley, Hoboken (2009)
21. Butun, I., Morgera, S.D., Shankar, R.: A survey of intrusion detection system in wireless sensor networks. IEEE Commun. Surv. Tutorials **16**(1), 266–282 (2014)
22. Simenthy, J.R., Vijayan, K.: Advanced intrusion detection system for wireless sensor networks. IJAREEIE **3**(3), 167–172 (2014)

23. Amita, G., Subir, H.: A survey on energy efficient intrusion detection in wireless sensor networks. JAISE **9**(2), 239–261 (2017)
24. Mitche, R., Chen, I.R.: A survey on intrusion detection in wireless sensor network applications. Comput. Commun. **42**, 1–23 (2014)
25. Singh, J., Thaper, V.: Intrusion detection system in wireless sensor networks. IJCSCE **1**, 2 (2012)
26. Kamaev, A., Finogeev, A.G., Finogeev, A.A., Parygin, D.S.: Journal of Physics: Conference Series, vol. 803(1) (2017)
27. Sathya, D., Krishneswari, K.: A Novel Cross Layer Rule Based Intrusion Detection System to Detect the Attacks Coming from Different Layers in WSN (2016). http://nopr.niscair.res.in/handle/123456789/34052
28. Yarvis, M., Kushalnagar, N., Singh, H., Rangarajan, A., Liu, Y., Singh, S.: Exploiting Heterogeneity in Sensor Networks, 5th edn., vol. 8. AK Press (2007)
29. Wang, X., Yoo, Y., Wang, Y., Agrawal, D.P.: Impact of Node Density and Sensing Range on Intrusion Detection in Wireless Sensor Networks, 6th edn., vol. 2. ECW Press (2006)
30. www.ijcsit.com; 1000projects.org; www.ijetr.org; www.ukessays.com; www.jpinfo.org; www.rroij.com; etd.ohiolink.edu; forums.havenworld.co.uk; www.ijrte.org; theglobaljournals.com

Conservation of Feature Sub-spaces Across Rootkit Sub-families

Prasenjit Das[✉]

Chitkara University, Baddi, Himachal Pradesh, India
Prasenjit.das@chitkarauniversity.edu.in

Abstract. Modern malware detection systems have largely relied on the definition of signatures to characterize malwares to their corresponding malware families. These signatures that characterize malware families are parts of codes and it is believed that families of malwares share commonalities in their signatures. We hypothesize that changes in these signatures generate newer sub-families of malwares. In the present work we have evaluated the signature conservation across two sub-families of rootkits. We have carried out our experiments to establish the fact that features in the rootkit family of malware are conserved. We report that our feature extraction yielded the accuracy of 84.17% using the Naïve Bayes classification algorithm. The results reported in this work reinforce our belief that there are subsets of independent features that discriminate between sub-families but not exhibiting any trend of conservation. We conclude that certain features (if not all) are preserved and discriminate between sub-families.

Keywords: Data mining · Malware · Rootkit · Classification · Clustering Bi-clustering

1 Introduction

Information that is available on the standalone as well as network systems is constantly under the threat of being attacked by software trying to malign the working of the system or steal the data. In general, most of the malwares (malicious software's) are commonly referred to as a virus. Considering the detrimental effects of these malicious software's, organization largely rely on intermediate intrusion detection systems to detect and block malwares from entering and causing damage. It should be noted that not all 'virus attacks' are caused by viruses, requiring intrusion detection systems to be constantly up-to-date with known malware families and their corresponding morphological changes as these malwares constantly evolve, rendering detection ineffective. Malwares are broadly categorized based on the criteria such as: the manner by which these malware enters into the system, the manner by which they execute, and by the associated damage that they can cause.

Reported by Microsoft^{TM}, there exists close to two hundred families of malwares. Some of the major malware families are *Computer Virus, Trojan horse, Worms, Spyware, rootkits.* Furthermore, as per this study conducted in 2006 by Microsoft, approximately 75% of the malware detected belong to 25 prominent families of

R. Sharma et al. (Eds.): ICAN 2017, CCIS 805, pp. 179–191, 2018.
https://doi.org/10.1007/978-981-13-0755-3_14

malwares and their variants [1]. This raises the question that if the variants of malwares are indeed related to common families, then they must share common behavior. This then logically implies that the code of the malwares must share something in common. A rootkit is a malware which is designed to hide the fact that a system has been infected by the malware by changing its important executable files required by the operating system to deliver the desired outcome. Rootkit is different from other malwares as this malware attacks the operating system files.

The motivation of this work is to explore relationships between rootkit malware sub-families, and more specifically we explore the feature subspace conservation. We hypothesize that there exists a relation between the subfamilies of rootkits namely HLLP and HLLW, and that these relationships are conserved across samples of both HLLP and HLLW. The following manuscript is structured as follows: Sect. 2 provides an overview of recent advances in malware detection techniques. Section 3 explains the feature extraction and feature selection along with the classification carried out. Our results and discussions are explained in Sect. 4, followed by conclusion in Sect. 5.

2 Related Work

Traditional malware detection techniques rely on a database of known signatures about malwares, and its ability to detect these signatures in real time. However, traditional method of signature detection is rendered ineffective against newer malwares, for example, in the case of zero day attacks. Furthermore it is stated that as the number of known viruses increased the size of signature database also increased resulting in increase of time taken for checking a file for malware signature [2]. According to David Perry, a computer specialist with Trend Micro™ which is a leading giant in the field of Antivirus, more than five thousand new malware samples are created in day which simply outnumbers the counter measures being developed. Earlier the malwares were only confined to standalone systems for which anti viruses were able to counter the effect of the malware. Unknown or new viruses easily escape the detection technique of signature based counter measures [3–12] which are results of the recent research works undertaken by various researchers. Based on the features of a particular malware family, classification and clustering of malware have been performed.

Detection of Trojans using techniques of data mining has been performed by Siddiqui et al. in [13]. The dataset used had 4722 Portable Executable consisting of 3000 Trojans and 1772 clean. For dimensionality reduction, Principal Component Analysis and Random Forest was applied on the dataset. Since the initial works of Nataraj et al. [14] and Kolter et al. [15], machine learning techniques have been used in many efforts to automatically classify unknown malware. For automated malware classification in [16, 17], discriminatory features were explored in 43 malicious file which led to the categorization of the files into respective malware families. SVM, kNN and the decision tree are several most popular classifiers used for malware classification problems [14, 18].

Clustering methods have also been applied on malware features to automatically identify the malware families based on the similarity of the features in [19–21]. Bit shred bi-clustering technique was used in [19] where owing to large no. of features that malware families have, the same were reduced using the feature hashing technique. Feature hashing helped in dimensionality reduction resulting in lower feature space. Similarly, clustering was used in the behavioral analysis of the malware based on premise of feature similarity to facilitate the grouping of malware samples into clusters [21]. The work used Locality Sensitive Hashing (LSH) [39], where the LSH algorithm computes similarity using the Jacard's coefficient and cosine similarity. Hierarchical clustering had been applied in [20] where the behaviors of the malware were labeled and the labels (referred to as fingerprints) were used to find the similarity between two malware samples. Edit distance had been used to cluster the similar fingerprints.

3 Methodology

Malwares belonging to same family exhibit certain features are common to all the instances of the said family. Clustering techniques group together the samples based on their feature similarity into clusters wherein each cluster has a high intra-cluster similarity and low inter cluster similarity. That is the in each subset (cluster) the samples are similar to one another and dissimilar to samples in other clusters. Clustering methods can be applied either to row or columns of a data matrix, separately. Bi-clustering methods on the other hand, perform clustering in two dimensions simultaneously.

The goal of this work is to employ bi-clustering which can identify subgroups of instances and subgroups of features by performing simultaneous clustering of both rows and columns instead of clustering these two separately. Bi-clustering algorithms assume one of the following situations: either there is only one bi-cluster in the data set consisting of rows and columns or there are K bi-clusters where K is number of bi-clusters we expect to find in the given data set. The K in the second approach is defined prior. Though there are algorithms in Bi-clustering which try to find only one bi-cluster in the data matrix, most of the algorithms [22–32] assume the existence of several bi-clusters in the data matrix. Clustering alone works like a black box revealing that the similar features form a cluster (so malwares are grouped since they have similar feature). Bi-clustering takes it one step further by telling us why are features similar (and hence grouped) by clustering features simultaneously.

In the present work, we have applied the data mining techniques to check the dynamic behavior of root kit family of malware to investigate that the features of rootkit instances are conserved across the classes of the rootkit.

The rest of the paper is arranged as per the following: Dataset used and feature extraction is explained in Sect. 3.1, Feature selection and ranking techniques applied on the data set in Sect. 3.2. Applying the supervised learning technique on the dataset to establish that the hierarchy of the features can be used to detect the malicious behavior of a binary executable and bi-clustering is explained in Sect. 3.3. Validations applied on the results obtained are explained in Sect. 3.4.

3.1 Dataset and Feature Extraction

The dataset used for the experiments has 2 classes of rootkit malware family namely HHLP and HLLW. It has in all 32 instances of malwares out which we have extracted 100508 attributes in a CSV file. The root kit dataset that has been used for experiments has been downloaded from the source [33]. Hexdump[1] tool has been used to extract the hexadecimal code from the binary rootkit file.

Feature extraction of the root kit data set has been accomplished using the n-gram analysis. An n-gram is a sequence of bytes of fixed or variable length, extracted from the hexadecimal dump of an executable program. Here, we are using n-grams (n = 5) as a general term for both overlapping and non-overlapping byte sequences. In the present work we have used the non-overlapping sequence of bytes.

For a given binary executable file of rootkit, UNIX hexdump utility is applied to convert binary executable into its equivalent hexadecimal number in form of a text file that is comma delimited (CSV file). Given the CSV file, the extraction of *n*-gram is carried out using the sliding window method without any overlap. We achieve this by applying the mid function provided to us in Microsoft Excel™. Here we select a particular substring from a given string. This conversion, converts each malware file to a set of features. Our next task is to represent each malware file by those features that are common to every other file in the datasets. We therefore resort to using MySQL™ as follows:

The resulting dataset generated is stored in a database where every file is described by its corresponding features. To bring in uniform representation of each malware file, we select only those features that are common to all malware files in the database. This is done by applying the left join on the tabular format of the features along with its class and taking its union thereof.

3.2 Feature Selection and Ranking

A feature selection algorithm is a combination of searching techniques of feature subset and evaluation which can be used to score the feature subset. Another purpose of performing a feature selection is that in a given dataset of malware there are a large number of features which are irrelevant in terms of their contribution towards the malicious behavior of the binary executable. These features need to be removed because in presence of these features the data mining algorithm to be applied on the dataset may not perform well. Feature weighting can be viewed as a generalization of feature selection. Feature weighting provides a better differentiation between features as it assigns a continuous valued weight for each of the feature. For the selected dataset of rootkits, in the present work, we have applied the feature selection techniques.

3.2.1 Information Gain

Information gain: This feature selection measure provides a ranking for each feature in the given data set. Information gain is a very effective metric in selecting features which is measured by calculating the entropy. Information gain can be defined as a

[1] http://avl.enemy.org/utils/hextools/.

measure of the effectiveness of an attribute (i.e. feature) in classifying the training data [34]. The training dataset when split on the basis of this attribute, information gain provides us a measurement of reduction of entropy that is expected after splitting the dataset. A feature is able to classify the data better, if the feature can reduce the entropy in the training data more effectively. Information Gain of a binary attribute A on a collection of examples S is given by

$$Gain(S,A) = Entropy(S) - \Sigma|S_V|/|S|Entropy(S_V) \tag{1}$$

where A is the set of all possible values for attribute A, and S_v is the subset of S for which attribute A has value v. In our case, each binary attribute has only two possible values (0, 1). Gain(S, A) tells us how much would be gained by branching on the attribute A. The attribute A contributing to the highest value of information gain is chosen as the splitting value. Entropy of subset S is computed using the following equation:

$$Entropy(S) = -\frac{p(s)}{n(s)+p(s)}log2\frac{p(s)}{n(s)+p(s)} - \frac{n(s)}{n(s)+p(s)}log2\frac{n(s)}{n(s)+p(s)} \tag{2}$$

where p(S) is the number of positive examples in S and n(S) is the total number of negative examples in S. The entropy of the subset S is measure of the disorder/variation in it.

3.2.2 Gain Ratio

Gain ratio (GR) is a modification of the information gain. This technique reduces its bias towards the features with larger values which is a limitation information gain. Gain ratio takes number and size of branches into account when choosing an attribute. It rectifies the information technique as it takes into consideration the intrinsic information generated by the split. Gain ratio is an extension of info gain which applies a kind of normalization to information gain using a split information value defined as

$$SplitInfo_A(V) = -\sum|S_V|/|S|Xlog_2(|S_V|/|S|) \tag{3}$$

This value represents the amount (potential) of information that is generated by splitting the training data set S into v partitions. It is in correspondence with the outcomes that we get after running the test on attributes of A. It is different from the information gain as the information gain measures the information with respect to classification obtained based on the same partitioning. The gain ratio is calculated using the formula:

$$GainRatio(A) = \frac{Gain(A)}{SplitInfoA(V)} \tag{4}$$

3.2.3 Relief

Relief algorithm is considered to be a feature subset selection algorithm at the time of pre-processing the dataset and uses the k-nearest neighbors (kNN) for a given feature. Relief uses instance based learning to assign a relevance weight to each feature. Each feature's weight reflects its ability to distinguish among the class values. Features are ranked by weight and those that exceed a user-specified threshold are selected to form the final subset. An attribute's weight is determined by how well the values if a particular attribute is able to differentiate the instances in the sample from its nearest hit and nearest miss. If an attribute is able to differentiate between the samples of different classes and has the same value for samples of same class, the attribute is assigned a high weight as this attribute is considered an important attribute which can distinguish amongst its class samples. Following shows the weight calculating formula used by

$$W_X = W_X - \frac{diff(X,R,H)^2}{m} + \frac{diff(X,R,M)^2}{m} \quad (5)$$

where W_X is the weight for attribute X, R is a randomly sampled instance, H is the nearest hit, M is the nearest miss, and m is the number of randomly sampled instances. The function diff calculates the difference between two instances for a given attribute. For nominal attributes it is defined as either 1 (the values are different) or 0 (the values are the same), while for continuous attributes the difference is the actual difference normalized to the interval [0, 1]. Dividing by m guarantees that all weights are in the interval [–1, 1].

3.3 Classification and Bi-Clustering

After selecting the features as per the ranking provided by the feature selection method, we have applied 2 classifications to investigate the relationship between the features. Naïve Bayes and Decision Tree have been applied to the selected feature sets. Naïve Bayes works under the assumption that the features are independent of each other. The weights assigned in the feature selection process are to find out if there is a relationship between the features with class labels. It is based on Bayes theorem of posterior probability. The classifier assigns the most likely class for the instance best described by its feature set. Following is the formula for predicting the class based on features using posterior probability:

$$P(X|C) = \Pi\, i = 1\, to\, n\, P(Xi|C) \quad (6)$$

where $X = \{X_1, \ldots, X_n\}$. is the feature vector and C is a class.

Decision Tree is a greedy Approach method which assumes that features are not independent of each other. C4.5 [35], and its predecessor, ID3 [36], are algorithms that summarize training data in the form of a decision tree. Decision trees are an effective method of supervised learning. The intent of applying decision tree is to partition the dataset into groups as similar as possible in terms of the class to be predicted. The outcome of the classifier is a tree where each leaf node is a predicted class and a non-leaf node is a test that has been performed under the certain conditions (represented by edge) of the tree.

Using logical rules, decision tree algorithms are a popular practice. This is due to two reasons. One is the execution speed and robustness. Another reason is that the outcome generated has explicit description that can be easily interpreted by the user. Both C4.5 and ID3 use a greedy approach to build a decision tree from training data. Information theoretic measure is used as a guide for the greedy approach applied. C4.5 selects the feature from the dataset which splits the instances of the malware into one class or the other. The criterion for splitting is the difference in entropy that results from selecting the feature for splitting the data. The feature with the highest difference in entropy is used as used to make the decision. ID3 has a limitation that it is sensitive towards features with large no. of values. It selects the best attribute based on entropy and information gain. C4.5 overcomes this limitation of the ID3 algorithm by selecting the splitting value based on Gain ratio which is not sensitive towards the features with large no. of values. Also C4.5 can be used on continuous data and can take care of the unknown (missing) values which are an improvement over the ID3 algorithm. C4.5 is able to overcome another limitation of ID3 that features of different weights can be used in C4.5 algorithm [37].

We have applied Bi-clustering method to establish if the features are conserved between the classes. We have used LAS algorithm for the same. LAS, developed by [38] are method designed to find bi-clusters with a large average as compared to the remaining dataset. Mathematically, let X be an m by n matrix of entries x_{ij} for i = 1,…, m and j = 1,…, n. Then, a bi-cluster U consists of a subset of rows A and columns B of X:

$$U = \{x_{ij}, : i \in A, j \in B\} \quad (7)$$

X is the sample which is in the form of row and columns (matrix) and x_{ij} are the feature values. The order of sample and feature values in not retained when we apply bi-clustering. In the present work the *i*'s represent the samples and *j*'s represent the features. A score function based on Bonferroni significance correction is calculated for each bi-cluster which trades off between the sub-matrix size and average value.

3.4 Validation

Once the model has been developed using the classification techniques Naïve Bayes and decision tree, we need to validate the same. A confusion matrix which shows the no. of True Positives (TP), True Negatives (TN), False Positives (FP) and False Negatives (FN) is a good measure of accuracy of a particular classifier. The following is an explanation of what each of these measures in confusion matrix man.

TP: Positive tuples correctly labeled correctly by the classifier.
TN: Negative tuples correctly labeled by the classifier.
FP: Negative tuples incorrectly labeled as positives by the classifier.
FN: Positive tuples incorrectly labeled as negatives by the classifier.

The validation criteria based on the above is calculated using accuracy, precision and recall measures for the classifier. The formulas are provided below:

$$Accuracy = \frac{TP + TN}{P} \tag{8}$$

$$Recall = \frac{TP}{P} \tag{9}$$

$$Precision = \frac{TP}{TP + FP} \tag{10}$$

where P is the number of positive tuples.

We have applied k (k = 10) fold cross validation to generate the confusion matrix. In k fold cross validation, the dataset D is divided into k subsets namely D_1, D_2, ….., Dk. Each of these subsets is approximately of equal size. Training and testing is performed k times. In iteration *i*, the subset D_i is reserved as the test set and the rest of the D_{k-1} subsets are collectively used for training the model. In this validation technique, each subset is used the same number of times for training and once for testing. Validation of the bi-clustering score is done using the Bonferroni significance correction. The Bonferroni correction is an adjustment made to P values when several dependent or independent statistical tests are being performed simultaneously on a single data set. To perform a Bonferroni correction, the critical P value (α) is divided by the number of comparisons being made. The Bonferroni correction is used to reduce the chances of obtaining false-positive results. So for the *k*-iteration of the Bonferroni correction the formula is

$$\alpha' = \frac{\alpha}{k} \tag{11}$$

where α is the level of significance (critical P value) test in *k*-th iteration and α' is the significance level of test we have to adapt to for each individual test.

4 Experiments and Results

The results that are obtained by applying different feature selection methods for generating model using supervised learning (classification) show that the accuracy of the model being generated either by Naïve Bayes or Decision Tree is same in terms of accuracy. Though the models generated using gain ratio and relief have lesser accuracy with Naïve Bayes than decision tree model. But overall the results obtained are almost similar. The dataset consists of 32 rootkit instances belonging to two malware classes (HLLP and HLLW). Feature extraction technique has yielded 100508 features with 17 instances of HLLP and 15 instances of HLLW.

The decision tree model works under the assumption that the attributes are related to each other. The measure of accuracy of the model obtained using the validation on training and testing data is an indicator of the existing relation between the attributes. The accuracies obtained from the experiments (applying various classifiers on the dataset) has been summarized in Tables 1, 2 and 3 respectively.

Table 1. Accuracy obtained from various classifiers.

Classifier	Accuracy
Information gain with Naïve Bayes	84.17%
Gain ratio with Naïve Bayes	77.50%
Relief with Naïve Bayes	74.17%
Information gain with decision tree	75.83%
Gain ratio with decision tree	75.83%
Relief with decision tree	68.33%

Table 2. Precision obtained from various classifiers.

Classifier	Precision
Information gain with Naïve Bayes	85.71%
Gain ratio with Naïve Bayes	75.00%
Relief with Naïve Bayes	81.82%
Information gain with decision tree	81.82%
Gain ratio with decision tree	81.82%
Relief with decision tree	77.78%

Table 3. Recall obtained from various classifiers.

Classifier	Recall
Information gain with Naïve Bayes	75.00%
Gain ratio with Naïve Bayes	75.00%
Relief with Naïve Bayes	50.00%
Information gain with decision tree	61.67%
Gain ratio with decision tree	61.67%
Relief with decision tree	40.00%

The following table (Table 4) shows the results of Bi-clustering LAS algorithm on data set:

The score in the above table signifies the true positives obtained from the experiments of Bi-clustering on all the features. That is of all the features present in the dataset the score value tells us the no. of features that are correctly identified in each of

Table 4. Bi-clusters score.

Bi-cluster No.	Size	Score
1	1 × 15732	12278.2714
2	2 × 15732	1214.0437
3	1 × 2083	145.6878
4	1 × 15732	75.5204
5	1 × 15732	0.8408

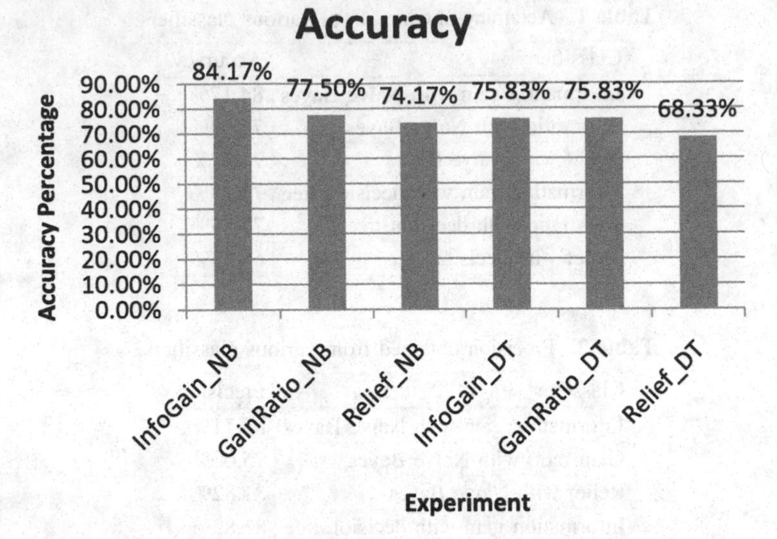

Fig. 1. Comparison of accuracies obtained

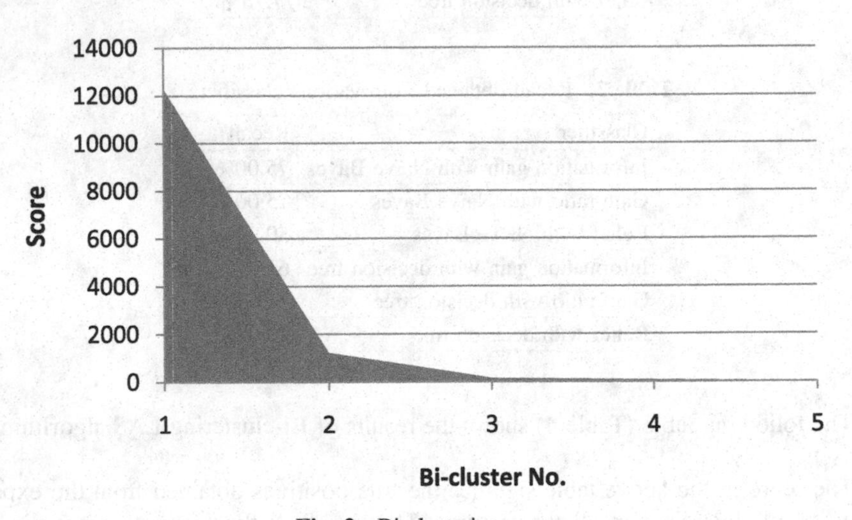

Fig. 2. Bi-clustering score

the cluster obtained. Also it captures the dynamic behavior of the malware that when the features are combined with different combinations they behave differently which is indicated by the different score shown above in Table 2. Figure 2 shows that the clusters obtained after applying the bi-clustering (LAS algorithm) has a high score indicating that the features are conserved in the sub-space leading to the high score of bi-clusters obtained.

5 Conclusion

The objective of the proposed work was gain a deeper understanding of the Rootkit family of malwares. We believed that there are relationships between signatures that require better elucidation for effective malware detection. Our specific aim of this work therefore was to create a data mining framework to effective characterize two sub-families of the Rootkit malwares – that relies of effective feature extraction, selection (or ranking of features). We then objectively evaluated the quantity of the features extracted using both the Naïve Bayes and Decision Tree (C4.5) algorithm. Our results from these experiments show that (refer Fig. 1) these features are effective in classifying samples of rootkit malwares, however, based on the highest accuracy of 84.17% using Naïve Bayes classifier reinforces the fact that the features are largely independent of each other.

We then applied the LAS bi-clustering, to prove that these features are not con-served across the two families of rootkits. The bi-clustering score shows that the score changes when the features are clustered together. Our scores reflect there are no relationships (conservation) among the features of the two sub-families of rootkit which can be explored with the inclusion of other families of rootkits.

Based on the classification accuracy of 84.17% obtained, and compared to the reported accuracy of Masud et al. [12], we understand that our features go against the hypothesis of the Hierarchical Feature Selection (HFS) algorithm. From this work we can conclude that our extracted features are effective in discriminating between sub-families of rootkits and it is worth exploring the inclusion of other malware families of interest. Furthermore, we believe that with the inclusion of newer families, and the inclusion of better classifiers like that of decision tree classifiers can explore the assumption that signatures of malware families are indeed dependent of each other.

References

1. Szor, P.: The Art of Computer Virus Research and Defense. Addison Wesley, Reading (2005)
2. Christodorescu, M., Jha, S.: Static analysis of executables to detect malicious patterns. In: Proceedings of the 12th USENIX Security Symposium (Security 2003), pp. 169–186. USENIX Association (2003)
3. McGraw, G., Morrisett, G.: Attacking malicious code: a report to the infosec research council. IEEE Soft. **17**(5), 33–41 (2000)
4. Golbeck, J., Hendler, J.: Reputation network analysis for email filtering. In: CEAS (2004)
5. Newman, M.E.J., Forrest, S., Balthrop, J.: Email networks and the spread of computer viruses. Phys. Rev. E **66**, 035101 (2002)
6. Schultz, M., Eskin, E., Zadok, E.: MEF: malicious email filter, a UNIX mail filter that detects malicious windows executables. In: USENIX Annual Technical Conference - FREENIX Track, June 2001
7. Masud, M.M., Khan, L., Thuraisingham, B.: Feature based techniques for auto-detection of novel email worms. In: The Eleventh Pacific-Asia Conference on Knowledge Discovery and Data Mining (PAKDD) (2007)

8. Singh, S., Estan, C., Varghese, G., Savage, S.: The Earlybird system for real-time detection of unknown worms. Technical report – cs 2003–0761, UCSD (2003)
9. Kim, H.A., Karp, B.: Autograph: toward automated, distributed worm signature detection. In: The Proceedings of the 13th Usenix Security Symposium (Security 2004), San Diego, CA, August 2004
10. Newsome, J., Karp, B., Song, D.: Polygraph: automatically generating signatures for polymorphic worms. In: Proceedings of the IEEE Symposium on Security and Privacy, May 2005
11. Schultz, M., Eskin, E., Zadok, E., Stolfo, S.: Data mining methods for detection of new malicious executables. In: Proceedings of IEEE Symposium on Security and Privacy, pp. 178–184 (2001)
12. Masud, M.M., Khan, L., Thuraisingham, B.: A hybrid model to detect malicious executables. In: Proceedings of 2007 IEEE International Conference on Communications, pp. 1443–1448. IEEE, June 2007
13. Siddiqui, M., Wang, M.C., Lee, J.: Detecting trojans using data mining techniques. In: Hussain, D.M.A., Rajput, A.Q.K., Chowdhry, B.S., Gee, Q. (eds.) IMTIC 2008. CCIS, vol. 20, pp. 400–411. Springer, Heidelberg (2008). https://doi.org/10.1007/978-3-540-89853-5_43
14. Nataraj, L., Yegneswaran, V., Porras, P., Zhang, J.: A comparative assessment of malware classification using binary texture analysis and dynamic analysis. In: ACM AISec 2011 (2011)
15. Kolter, J.Z., Maloof, M.A.: Learning to detect and classify malicious executables in the wild. J. Mach. Learn. Res. **7**, 2721–2744 (2006)
16. Perdisci, R., Lanzi, A., Lee, W.: Mcboost: boosting scalability in malware collection and analysis using statistical classification of executables. In: ACSAC 2008 (2008)
17. Rieck, K., Trinius, P., Willems, C., Holz, T.: Automatic analysis of malware behavior using machine learning. J. Comput. Secur. **19**(4), 639–668 (2011)
18. Rieck, K., Krueger, T., Dewald, A.: Cujo: efficient detection and prevention of drive-bydownload attacks. In: ACSAC 2010 (2010)
19. Jang, J., Brumley, D., Venkataraman, S.: Bitshred: feature hashing malware for scalable triage and semantic analysis. In: Proceedings of ACM CCS 2011 (2011)
20. Bailey, M., Oberheide, J., Andersen, J., Mao, Z.M., Jahanian, F., Nazario, J.: Automated classification and analysis of internet malware. In: Kruegel, C., Lippmann, R., Clark, A. (eds.) RAID 2007. LNCS, vol. 4637, pp. 178–197. Springer, Heidelberg (2007). https://doi.org/10.1007/978-3-540-74320-0_10
21. Bayer, U., Comparetti, P.M., Hlauschek, C., Kruegel, C., Kirda, E.: Scalable, behavior-based malware clustering. In: NDSS 2009 (2009)
22. Hartigan, J.A.: Direct clustering of a data matrix. J. Am. Statis. Assoc. **67**(337), 123–129 (1972)
23. Cheng, Y., Church, G.: Biclustering of expression data. In: International Conference on Intelligent Systems for Molecular Biology (ISMB), Department of Genetics, Harvard Medical School, Boston, MA 02115, USA, vol. 8, pp. 93–103 (1999)
24. Getz, G., Levine, E., Domany, E.: Coupled two-way clustering analysis of gene microarray data. Proc. Nat. Acad. Sci. **97**(22), 12079–12084 (2000)
25. Califano, A., Stolovitzky, G., Tu, Y.: Analysis of gene expression microarrays for phenotype classification. In: Proceedings of International Conference on Intelligent Systems for Molecular Biology (ISMB), vol. 8, pp. 75–85 (2000)
26. Lazzeroni, L., Owen, A.: Plaid models for gene expression data. Statistica Sinica **12**, 61–86 (2002)
27. Segal, E., et al.: Rich probabilistic models for gene expression. Bioinformatics, **17** Suppl 1 (1), S243–S252 (2001)

28. Tang, C., et al.: Interrelated two-way clustering: an unsupervised approach for gene expression data analysis. In: Proceedings - 2nd Annual IEEE International Symposium on Bioinformatics and Bioengineering, BIBE 2001, pp. 41–48 (2001)
29. Yang, J., et al.: Delta-clusters: capturing subspace correlation in a large data set. In: Proceedings of 18th International Conference on Data Engineering, p. 12 (2002)
30. Kluger, Y., et al.: Spectral biclustering of microarray data: coclustering genes and conditions. Genome Res. **13**(4), 703–716 (2003)
31. Segal, E., Battle, A., Koller, D.: Decomposing gene expression into cellular processes. In: Pacific Symposium on Biocomputing, pp. 89–100 (2003)
32. Liu, J., Wang, W.: OP-cluster: clustering by tendency in high dimensional space. In: Proceedings of Third IEEE International Conference on Data Mining, pp. 187–194 (2003)
33. https://vxheaven.org/
34. Mitchell, T.M.: Machine Learning. McGraw-Hill, Maidenhead (1997)
35. Quinlan, J.R.: Programs for machine learning. Mach. Learn. **240**, 302 (1993)
36. Quinlan, J.R.: Induction of decision trees. Mach. Learn. **1**(1), 81–106 (1986)
37. Hssina, B., Merbouha, A., Ezzikouri, H., Erritali, M.: A comparative study of decision tree ID3 and C4.5. Int. J. Adv. Comput. Sci. Appl. **4**(2) (2014)
38. Shabalin, A.A., Weigman, V.J., Perou, C.M., Nobel, A.B.: Finding large average submatrices in high dimensional data. Ann. Appl. Statis. **3**, 985–1012 (2009)
39. Indyk, P., Motwani, R.: Approximate nearest neighbors: towards removing the curse of dimensionality. In: Proceedings of 30th STOC, pp. 604–613 (1998)

Gini Coefficient Based Wealth Distribution in the Bitcoin Network: A Case Study

Manas Gupta[1(✉)] and Parth Gupta[2]

[1] Department of Computing, National University, Singapore, Singapore
manas90mg@gmail.com
[2] Department of Computer Science Engineering, Chitkara University, Rajpura, Punjab, India
parthguptapg19@gmail.com

Abstract. Bitcoin has gained widespread attention globally in 2013 and is the first online currency based on a peer to peer network without any central authority or third parties. Its market capitalization reached US$ 8.5 billion in December 2013. However, despite its popularity some issues like network security (thefts), anonymity (privacy) and wealth distribution (inequality) have plagued it. Of considerable importance is the last issue of unequal wealth distribution as it may create a huge socio-economic burden for the society. A group of researchers estimated that the GINI coefficient for the network was at an all time high of 0.985 in Jan 2013 and that the rich were getting richer as the network grew. In the present work it has been strived to determine how the GINI actually increases or decreased depending upon the wealth distribution. For doing this a raw transaction of data of more than 36 million transactions has been sourced and a list of all users and their wealth in the network has been computed. The final results are very alarming as GINI has increased to 0.997 by the end of 2013 and the market share of top 10 holders alone has reached 6.6% of the entire market. Therefore, the rich have actually got richer and steps should be taken to curb such a wealth accumulation model in the network.

Keywords: Bitcoin protocol · Gini coefficient · Wealth distribution
Bitcoin wallet

1 Introduction

Bitcoin has gained widespread attention globally in 2013. Most people have now heard about Bitcoins and know that it is a digital currency. This has mainly been due to the huge fluctuations in the trading price of Bitcoins caused by financial turmoils like the Cyprus crisis from Feb-May 2013. During that time Bitcoin prices jumped from USD 20 per BTC (Bitcoin) to above USD 260 per BTC1over a span of two months triggering extensive media coverage [1]. Current prices are around USD 220 per BTC and gaining momentum rapidly over the past few months. However, the greatness of Bitcoin does not stem from its meteoric price rise but from its inherent economic and technical robustness which will be elaborate more in this paper.

R. Sharma et al. (Eds.): ICAN 2017, CCIS 805, pp. 192–202, 2018.
https://doi.org/10.1007/978-981-13-0755-3_15

Bitcoin is also the first online currency based on a Peer to Peer network without the use of any central authority or 3rd parties. It is completely independent of any single individual or institution running the protocol and relies on the nodes in the network for the smooth functioning of the network. It has grown very rapidly over the last 5 years and the current circulation value of bitcoins is greater than USD 8 billion [2–4].

2 Literature Review

Bitcoin has grown a lot since 2009. Along the way it has encountered various problems including but not limited to price fluctuations, regulatory challenges, illegal usage, cyber attacks and rival currencies. Also, with the passage of time, the academia has begun identifying and recommending solutions to some of these problems. Doing a holistic literature review identified some key areas where research is currently being done.

Network Analysis. Network analysis entails studying the Blockchain which is publicly available and creating a network graph based on that. Since, the Blockchain contains transactions, this first graph that is created is a transactions graph. Some studies have processed this graph further to yield a user graph which combines various transactions done by a single user. However, this is a non trivial step and various assumptions are required to be made in order to do the processing. These graphs can be studied in various ways depending on what results are required by the writers. Some key ones are-

Anonymity. One of the first network studies on Bitcoin was undertaken by Martin Harrigan and Fergal Reid to study the degree of anonymity [3] in the Bitcoin network.

Since, anonymity is not one of the design considerations of Bitcoin, user privacy is a critical area that is of concern to many users. This paper [3] specifically tried to target a case of theft where bitcoins worth half a million dollars were stolen and identified users that might be related to the theft. The main methodology used to derive the user graph from the transaction graph was that, when a transaction with multiple inputs happens, it can be realized that all the inputs belong to the same user. This property was then used in all transactions to reduce the graph to the user graph. Once, the authors get the user graph, they then focus on the theft, and by looking at users linked to the theft transaction indicate that one of the popular hacker groups called Lulz Security might be behind the attack.

Anonymity is not a key design consideration of the Bitcoin protocol. Although, users do not reference any personal identification information in any bitcoin transactions or protocols, still the acutely connected nature of the network means that any identification that can be linked to a user node in the network could lead to the exposure of a large number of connected nodes. For instance, many organizations like Wikileaks had put up their public keys online for anyone who wanted to donate to the organization. With this piece of information alone, one could easily trace the entire financial history of Wikileaks on the Bitcoin network. Even though they might try to bounce off their bitcoins from one address to another to increase anonymity, these transactions are very apparent once we refer to the user graph and can easily be unfolded.

Network Statistics. A second approach to network analysis was to look at the network itself and try to study interesting habits or properties of the network. Ron and Shamir [5] tried to find interesting network usage statistics pertaining to the Bitcoin network. However, the statistics are worth noting. For instance, they concluded that 60% of all bitcoins are "old" bitcoins that had not been transacted for a few months and are effectively out of circulation. There was potential to do similar graph based studies on the Bitcoin network and explore in more detail other aspects about the network usage as well as user behavior which is important in predicting how the network will grow and function over time.

Mining. Another interesting area about bitcoins is Bitcoin mining. Mining, as mentioned previously, is the act of validating a block of transactions and being rewarded if you are the first one to do that. As lucrative as it may sound, it is very resource intensive and needs a lot of computing power to be able to work. A new concept in this area is the act of creating mining pools where users devote computing power on a partial basis and get compensated for their devotion on a partial basis as well. Meni Rosenfeld has undertaken a study of such mining pools [6] and has tried to identify pitfalls with current reward systems for such pools. One pitfall is called pool hopping whereby users can get a skewed reward curve by joining the pool during times of high turnover and leaving it during low turnover periods. Rosenfeld has suggested a range of reward systems to solve the problem of hopping as well which seem feasible and robust.

Other Applications of Bitcoin. One of the key aspects of the Bitcoin system is its power to use peer to peer decentralized networks to validate and timestamp transactions. Some applications of this protocol can be used to solve problems in other areas as well. One such example is Commit Coin [7]. Commit coin builds on bitcoins and instead of just doing currency transactions allows a network to timestamp and validate any commitment or action. The example discussed in the paper is of applying it to election results whereby the votes are time stamped. Any tampering of votes once the vote is cast can be detected and rejected by the system. This is something promising however, in order for commit coin to take off in the real world the practicality of the solution needs to be considered as well. Firstly, the technology infrastructure required to support commit coin on a national basis can only be afforded by a select group of rich nations which leaves out the poor nations which are also the most susceptible to unfair elections. Secondly, it is not a fool proof method and tampering can still be done if the pre-election commitments are changed. Thus, more work needs to be done in order to make it financially and technically feasible.

To understand the current state of the network, many issues came up during the review including Anonymity, Mining, Privacy, Liquidity and Wealth Inequality etc. The most pressing of these challenges was understood to be the wealth distribution in the network. This is because wealth was found to be heavily centralized in the hands of a few users who own disproportionate amounts of bitcoins. The standardized measure for wealth distribution is the GINI coefficient which was found to be very high by a few studies.

One particular study which tackled this problem specifically was conducted by a group of data scientists from the Eotvos Lorand University, Budapest, Hungary [6].

They found that the GINI coefficient of the network as of Jan 2013 was at 0.985, which is supposed to be very high as the maximum allowed GINI coefficient is 1. They further found that sublinear preferential attachment drives the wealth accumulation in the network. In lay man terms, this means that the rich are getting richer. They predicted that this is how the network will grow in 2013 as well and that this trend has become the default way for the network to grow over time. If this is infact the case, then it poses a threat to the sustainability of the network as a few people cannot be relied on to drive the entire network worth billions of dollars.

Given the results of the study above, it seems to be a very pertinent issue to know whether sublinear preferential attachment actually happened in 2013. However, during the literature survey no study was found that measured and analyzed the wealth distribution characteristics in the network in 2013. Furthermore, some interesting catalysts took place in 2013. During the period from October 2013 to December 2013, the value of bitcoins jumped 10 fold from around USD 100 in October to more than USD 1000 in December 2013. This was based on the back of strong demand from investors around the world especially those in China.

Given such a huge increase in the user base of bitcoins, it can be intuitively conducted that wealth should have become less concentrated as the wealth spread over a larger base of users. This would be against what was predicted by the Hungarian scientists and would have actually led to the reduction of the GINI coefficient. Again, no study was found that analyzed this situation. Based on the two drivers mentioned above, a study of the bitcoin wealth characteristics in 2013 seemed very relevant and important. It is very shocking is that the Gini coefficient for the entire bitcoin network on 1 Jan 2013 was 0.985. The highest possible value for Gini coefficient is 1 and keeping this in view, 0.985 is a very high figure. Assuming 100 bitcoin are to be distributed among 100 users, GINI will be 0 when each user gets 1 BTC as shown in Fig. 1(a). On the other hand, if 100 bitcoin are to be distributed among 100 users, GINI will be 0.985 when 3 users receive 20, 30 and 50 BTC respectively as shown in Fig. 1(b).

As can be seen clearly from the Fig. 1, the discrepancy in bitcoin distribution among its users is huge. And this is an area which needs much more research.

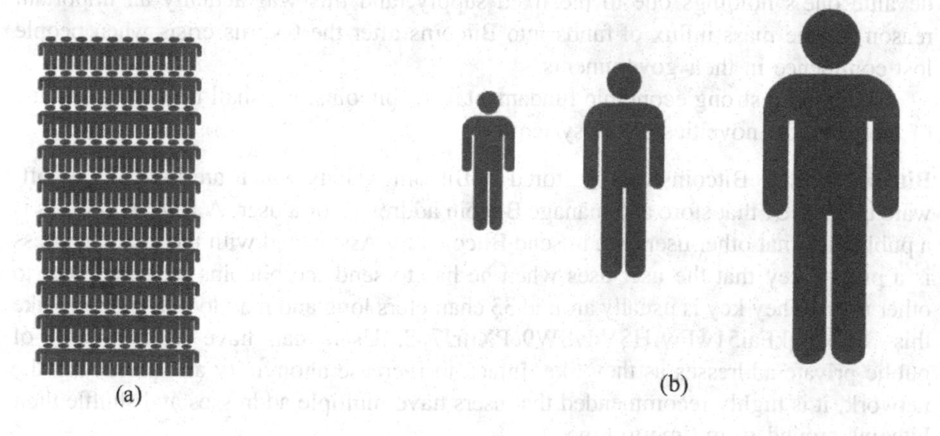

Fig. 1. Visualization of (a) GINI = 0, (b) GINI = 0.985

3 Bitcoin Protocol and Working of Bitcoin Network

In this section the concept of bitcoin protocol and working of the bitcoin protocol is elaborated. Virtual currencies as a concept has been around for many decades and the earliest conceptualizations of them arose in the 1980's. Yet no other crypto currency has taken off like Bitcoin. This does lead the question of why bitcoin has become so popular. Is it just media hype or does the protocol actually have something fundamentally different about itself. The bitcoin draws its huge strength and popularity from the core decentralized protocol that no single person or organization controls or can influence [5, 6]. This section will discuss about some of the main aspects of the protocol which include the nature of a bitcoin, the bitcoin wallet, transactions, blockchain and mining.

Bitcoin is a digital crypto currency facilitating transactions in the online world. It is a decentralized currency and does not have a single authority or server administering the currency. It is a peer to peer run network with the network nodes validating all the transactions happening in the network [8]. The idea of Bitcoin was floated by Satoshi Nakamoto [1], believed by many to be a pseudonym for a person or a group of persons in 2008. The first Bitcoin was generated on January 3, 2009 and led to the starting of the official block chain. Bitcoin is an open source software managed by the non profit Bitcoin foundation.

One of the key aspects differentiating Bitcoins from any other virtual or physical currency is its decentralized nature. This means that no single person, authority or organization controls or maintains the network. It cuts out the roles of intermediaries like banks or organizations like Paypal in the online world and governments and central banks in the physical world. Therefore, it allows a user to transact with anyone anywhere in the world without the need/interference of any third party.

Another interesting aspect of Bitcoins is that the money supply of bitcoins is fixed at 21 million BTC. This means that once all 21 million BTCs have been generated, no more bitcoins can be newly generated. This is in contrast to the real world where governments can print money at their discretion and cause a sudden drop in the purchasing power of the money held by citizens. Bitcoins, by design can thus never devalue one's holdings due to the fixed supply, and this was actually an important reason for the mass influx of funds into Bitcoins after the Cyprus crisis when people lost confidence in their governments.

Given these strong economic fundamentals of bitcoins, we shall now look at some of the technical novelties of the system.

Bitcoin Wallet. Bitcoins can be stored in Bitcoin wallets which are basically a software client GUI that store and manage Bitcoin addresses of a user. A Bitcoin address is a public key that other users use to send Bitcoins to. Associated with the public address is a private key that the user uses when he has to send the bitcoins in his address to other users. They key is usually around 33 characters long and may look something like this 3BTChqkFai51wFwrHSVdvSW9cPXifrJ7jC. Users can have as many set of public-private addresses as they like. Infact, to increase anonymity and privacy in the network, it is highly recommended that users have multiple addresses and shuffle their bitcoins around from time to time.

Transactions. Conducting a transaction entails the sender sending some bitcoins to the public key of the receiver. The sender digitally signs the transaction that can be later verified by the receiver. Each transaction has an input and an output. The input is either the reward for a newly found block or an output from a previous transaction transferring a certain number of bitcoins to the sender. The output states the address and amount to be sent to the receiver as well as the address that any change from the transaction needs to be sent to. Additionally senders can include a small transaction fee as well that gives an incentive for the nodes in the network to process this transaction before any non fee paying transactions. Another interesting feature is that each transaction can have multiple input and multiple outputs, thus offering great flexibility beyond that achievable in fiat cash currencies of today (Fig. 2).

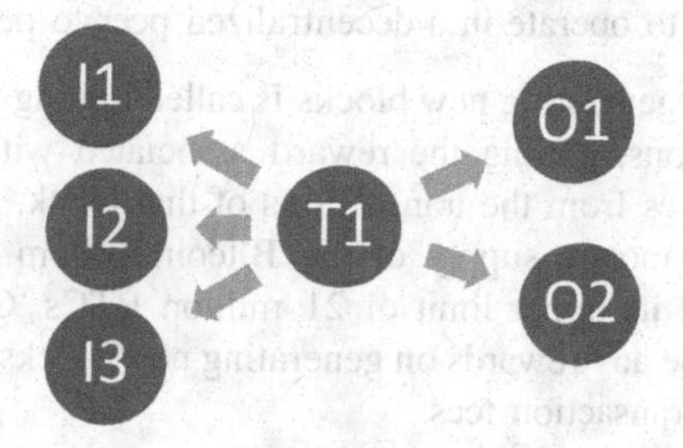

Fig. 2. Schematic of a typical bitcoin transaction

The transaction above titled T1 has 3 inputs I1, I2 and I3 and leads to 2 outputs O1 and O2. A typical transaction can have as many inputs and outputs as desired. However, there must be a minimum of 1 input and 1 output for a transaction to be valid. The outputs of one transaction form the inputs for the subsequent transaction.

Blockchain. Once the sender broadcasts the transaction to the node, multiple transactions are clubbed together in a block by the nodes working to solve/generate a new block.

Solving the block basically entails finding a solution to a cryptographic hash function. Whichever node or 'miner' is able to find the solution first is credited with generating the new block and is rewarded a certain number of Bitcoins. Currently, this reward is set to 25 BTCs. The newly generated block also has a reference to the last block generated. Thus, as new blocks are added, a chain of blocks is created, leading to it being called the Blockchain. There can only be 1 Blockchain in the network, and this Blockchain started with Nakamoto creating the first block in January 2009. How many blocks are generated is also controlled by the network and currently one block is generated every 10 min on average.

The Blockchain is one of the most important attributes of the Bitcoin system and one of its key differentiators from other proposed or existing crypto-currencies. It is required to ensure the decentralization, security and safety of the entire network (Fig. 3).

Fig. 3. Schematic of the blockchain

The blockchain is the series of all blocks ever created connected in order of creation. The blocks can have any amount of transactions inside them and are currently created at the rate of 1 block per 10 min. The rate of creation can vary and is not a fixed quantity. Notice, the last block in this chain is block 294691 which was created on 7 April, 2014. Blockchain is one of the major pillars of the protocol and is publicly known for the network to operate in a decentralized peer to peer manner.

Mining. The process of generating new blocks is called mining and the miner nodes do it because of two reasons, getting the reward associated with the new blocks and collecting transaction fees from the transactions of that block. It is also a way for the system to increase the money supply of the Bitcoin system in a gradual regulated manner until it reached its upper limit of 21 million BTCs. Once the upper limit is reached, there will not be any rewards on generating new blocks, and the only incentive for miners will be the transaction fees.

4 Gini Coefficient Analysis

Having arrived at the user and their associated wealth data in the last section, the final section of the analysis focuses on calculating the GINI coefficient for all the bitcoin users and then studying the results obtained. The following sections, will first discuss what the GINI coefficient is and how it is calculated. This paper will then discuss the key challenges faced in doing this, followed by the last section of listing and discussing the results obtained.

Gini Coefficient. The main goal for the statistical analysis is to prove or disprove the hypothesis of whether wealth concentration has increased in the system or not. This is similar to the concept of the Gini Coefficient, which is used to determine wealth distribution in a country. For instance, if Gini Coefficient is 0, then wealth is equally distributed among all citizens whereas if the coefficient is 1 then the wealth is concentrated and held by 1 person only. Hence, the lower the Gini coefficient, the better it is for a country.

The Lorenz curve is the underlying function used in the calculation of Gini. It is a cumulative distribution curve of the wealth of a country. As can be seen from the figure below, the number of people is plotted on the x-axis and their cumulative incomes/wealth on the y-axis (Fig. 4).

The Lorenz curve is a cumulative distribution curve which on the x-axis shows wealth held by each member and on the y-axis shows the total cumulative wealth per member. The shape of the Lorenz curve, i.e., the steepness of the curve determines how high the GINI coefficient will be.

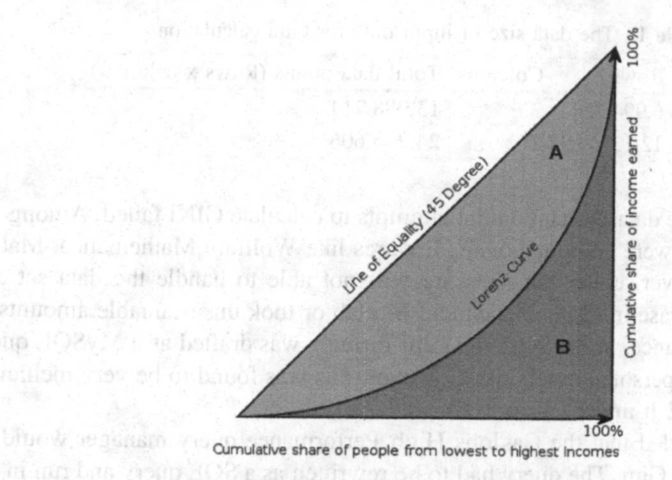

Fig. 4. Lorenz curve & cumulative wealth distribution

Gini Coeffcient *G*, is then given by

$$G = \frac{\text{Area } A}{\text{Area } (A + B)} \tag{1}$$

For this case, since the wealth distribution of the population (bitcoin users) is known, the Gini coefficient is calculated as-

$$G = \frac{1}{n}\left(n + 1 - 2\left(\frac{\sum_{i=1}^{n}(n+1-i)y_i}{\sum_{i=1}^{n} y_i}\right)\right) \tag{2}$$

where n is the number of bitcoin users in the network and y_i is the number of bitcoins held by user i.

The Gini coefficient and the Lorenz curve are the most standard and robust way to calculate wealth concentration and have been adopted by the OECD countries as well as the main measure of concentration analysis. We will adopt a similar methodology to find the Gini coefficients for the bitcoin network in May 2013 and Dec 2013. Such a thorough statistical analysis of the bitcoin network has never been done for the latest bitcoin data before and would be a pioneering effort in ascertaining the wealth distribution characteristics of the network as well as the evolution of the network over time.

Key Challenges and Approach Used. One of the biggest challenges in computing GINI was the size of data being handled. Since, we are calculating GINI for two time points in 2013 there are two data sets that were being operated on. The first dataset was that of May 2013 and the second that of December 2013. The table below summarized the data points being handled (Table 1).

Table 1. The data size of input data for Gini calculation.

Time point	Rows	Columns	Total data points (Rows x columns)
May 2013	6,994,357	2	13,988,714
Dec 2013	12,137,803	2	24,275,606

With such a huge dataset many initial attempts to calculate GINI failed. Among the approaches adopted were computational softwares like Wolfram Mathematica, Matlab and MySQL. However, either the software was not able to handle the data set and crashed as was the case in Mathematica and Matlab or took unsustainable amounts of time to run. For instance, in MySQL, the Gini formula was drafted as a MySQL query and was run on my personal notebook. However, this was found to be very inefficient as the query took **42 h and 51 min** to finish.

It was then decided that the CasJobs High Performance query manager would be used to calculate the Gini. The query had to be rewritten as a SQL query and run in the CasJobs interface which is much faster due to cloud computing and high performance servers.

In bitcoin network, the wealth is denoted in Satoshi where 1 BTC = 108 Satoshi. This forms as the input to the Gini query.

5 Results and Discussion

The obtained results are very informative and strongly in line with the initial hypothesis. They at the same time raise critical concerns about the Bitcoin network as a whole and whether bitcoin can stay strong in the future. In this section the further details about the various results obtained are discussed.

In this paper, the transactions are clubbed into blocks. The block number indicates how many blocks of transactions have happened so far in all and is a continuous chain. In the comparison time cycle considered in this paper, 42,443 new blocks were created.

Since, new bitcoins are created with the creation of every block, the total bitcoins in circulation increase with each new block. The net increase in bitcoin circulation was 1,061,075 BTC in this analysis or 25 BTC per new block created.

Also, there are the number of addresses or "virtual wallets" in circulation. It was observed that people created an astounding 11,532,431 new wallets. This is a very huge figure and indicates how the popularity of bitcoin rose tremendously during the analysis period as compared to the earlier existing 13,086,528 wallets, which had taken about 53 months to be created.

Whenever money goes from one wallet to another, it is categorized as a transaction. During the analysis, the transaction increased by 12,694,186.

The Gini coefficient is the international standard for measuring wealth distribution in a given population. The coefficient ranges from 0 to 1.0 implies that wealth is distributed equally among every citizen whereas 1 denotes that 1 person owns all the wealth in the entire population. The coefficient was 0.995 in year 2013. This is very high and clearly shows that bitcoins were very unequally distributed among the approximate 7 million population of bitcoin users in year 2013.

The above clearly shows the very unhealthy state of wealth distribution that the bitcoin network was in year 2013.

The Fig. 5 shows the wealth owned by the top 10 richest users in the bitcoin network as of year 2013. The x axis show the users in descending order of wealth and the y axis denotes their wealth in USD. The top ten users own more than 6.4% of the entire value of bitcoins.

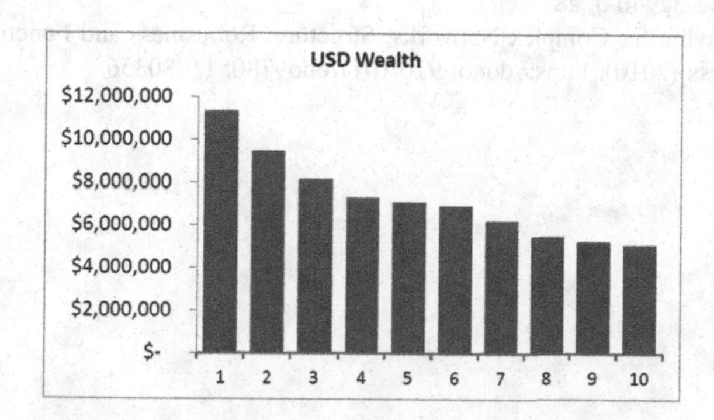

Fig. 5. Snapshot of the top 10 richest bitcoin users

6 Conclusion

In this paper the bitcoin network and its functioning had been elaborated. Since, the Bitcoin system is still in its preliminary state, a full understanding of the network, specifically with regards to user behavior needs to be study more. A wealth distribution study of the network by looking at user and transaction data had been undertaken, thereby understanding the dynamics and health of the network better.

However the analysis of GINI coefficient shows that the rich are getting richer, i.e., preferential attachment governs wealth accumulation in the bitcoin network. This is also a validation of the predictions of the Hungarian researchers and establishes a predominant network wealth growth model for bitcoins.

References

1. Nakamoto, S.: Bitcoin: a peer-to-peer electronic cash system. www.bitcoin.org
2. Kuzuno, H., Karam, C.: Blockchain explorer: an analytical process and investigation environment for bitcoin. In: 2017 APWG Symposium on Electronic Crime Research (eCrime), pp. 9–16 (2017)
3. Reid, F., Harrigan, M.: An analysis of anonymity in the bitcoin system. In: 2011 IEEE International Conference on Privacy, Security, Risk, and Trust, and IEEE International Conference on Social Computing, pp. 1318–1326 (2011)
4. Kondor, D., Pósfai, M., Csabai, I., Vattay, G.: Do the rich get richer? An empirical analysis of the BitCoin transaction network. Phys. Soc. (2014). https://doi.org/10.1371/journal.pone.0086197

5. Ron, D., Shamir, A.: Quantitative analysis of the full bitcoin transaction graph. In: Sadeghi, A.-R. (ed.) FC 2013. LNCS, vol. 7859, pp. 6–24. Springer, Heidelberg (2013). https://doi.org/10.1007/978-3-642-39884-1_2
6. Rosenfeld, M.: Analysis of Bitcoin Pooled Mining Reward Systems (2011). arXiv preprint arXiv:1112.4980
7. Clark, J., Essex, A.: CommitCoin: carbon dating commitments with bitcoin. In: Keromytis, A.D. (ed.) FC 2012. LNCS, vol. 7397, pp. 390–398. Springer, Heidelberg (2012). https://doi.org/10.1007/978-3-642-32946-3_28
8. Cohen, R., Havlin, S.: Complex Networks. Structure, Robustness and Function. Cambridge University Press (2010). https://doi.org/10.1017/cbo9780511780356

A Cipher Based Acknowledgement Approach for Securing MANETs by Mitigating Black and Gray Hole Attacks

Neha, Gurjot Kaur Walia(✉), and Harminder Kaur(✉)

Guru Nanak Dev Engineering College, Ludhiana, India
lakhanpalneha@yahoo.co.in, gurjotwalia@yahoo.com,
harminderkaur@gndec.ac.in

Abstract. Mobile indicates moving devices and Ad Hoc indicates temporary structure so mobile ad hoc network is a kind of temporary structure less networks. MANET is a collection of independent networks and this network is a non centralized, temporary mesh of continuously moving random nodes where each device in network works as a router and route packets to share data with each other. Due to its continuously evolving topology, limited resource and low safety, this is endangered to different attacks and black hole attack is one of them. To overcome this problem in MANET, a malicious node detection and encryption based approach has been used. In the proposed work reverse tracking mechanism has been used for detection of attacker nodes, that degrades performance of the network and a cipher based mechanism has been proposed to mitigate the multiple black hole and gray hole attacks, which yielded better results than other methods in terms of throughput, PDR, delay and network overhead.

Keywords: MANET · Security · Cipher text · Black hole attack
Gray hole attack · Nodes · Routing

1 Introduction

Due to large range of applications like in military, disaster areas etc. current years, have observed a great improvement in design of Mobile Ad-hoc Network technologies. Mobile ad-hoc network is self-architecture based, wireless network with stringent resource constraints, and rapidly changing topologies. In MANET, each and every intermediate node works as a router and route their packets. The network is ad-hoc; all device in network collection of mobile device. In mobile ad-hoc network device can share information with each other without any centralized access point [1]. Mobile devices are always in moving state and are connected through various wireless links. This network has capability to incorporate itself into large area network (LAN) [2].

Types of MANET are:

R. Sharma et al. (Eds.): ICAN 2017, CCIS 805, pp. 203–213, 2018.
https://doi.org/10.1007/978-981-13-0755-3_16

1.1 Vehicular Ad-Hoc Networks

In this Ad-Hoc network vehicles with wireless networking capabilities build a network. In wireless communication, equipment is rest interior of vehicles which is moving on the road for improving accessibility to other vehicles, to form a mesh for exchange information [3].

1.2 Intelligent Vehicular Ad-Hoc Networks

Vehicular Ad-Hoc Network which communicates through Wi-Max IEEE 802.16 and Wi-Fi 802.11 is known as InVANET. The prior objective of developing InVANET is to stop vehicle collision and ensure passengers safety. InVANET also aids drivers to keep secure distance among vehicles and also assist them and to determine speed of other vehicles. InVANET applications are also used in military for communication [4].

1.3 Internet Based Mobile Ad-Hoc Networks

iMANET supports internet protocols and making routes automatically [5].

The main aim of routing protocols in network is to generate shortest path between sender and receiver with low overhead and bandwidth consumption to ensure messages are convey in a timely manner. These can be divided into three categories:

Proactive Routing Protocol: In this protocol every node maintains tables that represent entire topology network. These tables maintain latest routing information from each node [7]. To maintain the current routing information, topology information regularly exchanged between the nodes, resulting in high overhead on network. In proactive routing, routes are always available on request. In proactive routing, devices are always share their routing messages with next node. It is necessary for every node to prepare the records of adjacent, approachable nodes with hop counts. It is an active routing environment as routes are always being available on demand and routing tables are updated every time. When topology change, original path becomes invalid and each device on the gathered information collected currently, change their table to establish new path. This continuous updating of routing table results in wastage of bandwidth and power of node. This routing protocols also termed as table driven protocols. In proactive routing, there is no delay for route determination, but at same time maintaining of route tables consume bandwidth.

Reactive Routing Protocol: This protocol is also termed as on-demand routing protocol i.e. routes are determined only when communication is needed to be done. Sender has to establish the route and transmit and receive the packet when needed. These protocols do not maintain routing information on all nodes [8]. Routing information is gathered only when it is required, and route discovery depends on sending route request message. The main convenience of this protocol is usage of a less bandwidth, and the limitation is that there is delay in transfer of the date as every node that sends packets, cannot rapidly discover path as routes are not always available. The path discovery steps create delays, and the average delay time is high in this type of protocol [9]. Example- AODV.

Hybrid routing protocol: This protocol uses both proactive and reactive routing protocol to overcome limitations of these protocols. Hybrid protocol is a adjustment among proactive and reactive protocols. Table driven protocols have high network overhead and low delay as of on-demand protocols have low overhead and high delay [10]. Thus a Hybrid protocol overcomes the limitations of both protocols [11]. It utilize the on demand characters of reactive protocol and the table driven characters of proactive protocol thus to avoid delay and network overhead issues [12]. Example - ZRP.

Further paper is organized as: Sect. 2 discusses the methods to be used to detect malicious nodes. Section 3, discusses various parameters used for performance evaluation of proposed method. Finally, the paper is concluded in Sect. 4.

2 Methdology

In the MANET, nodes start communicating from source to destination by using the multi-hoping process. In the process of data transmission request message that contain hello message has been broadcast to the nodes that are available in the transmission range. In the broadcasting of the message request message has been transmitted under IPV6 format. In the process of MANET various routing protocols have been used that are reactive or on demand has been used for data transmission.

In this work multi-hop approach is used for data transmission from source to destination using different routing protocol. In the MANET various attackers performs the attacks for degradation of the performance of the entire network. In this network different attacks like grey hole and black hole attack has been performed on the network.

Black hole attacks occurs by selfish node that just drops the packets. It occurs on network layer and is active type of attack. Malicious nodes does not pass any packet to the destination instead it drops all the packets. In Black hole attack, malicious node declares to have a reliable shortest path, and after the path is formed it drag packets without passing to other node. As mentioned above, attacker node may be leak or consumes the data packets. As a result, the network function is undergo extremely from this problem. The major impact is PDR decreases sharply.

Detection phase

In the process of proposed work encryption has been done that using public and private key based encryption. Data packets and route packets have been encrypted using AES encryption approach. Using encryption approach data integrity has been enabled that prevents various information leakages under different attacks. In this work data integrity and malicious node detection has been done on the basis of encryption approach and reverse tracking mechanism.

3 Results

On the basis of simulation parameters as shown in Table 1 network has been initialized and the routing protocol has been used for transmission of information under different simulation scenarios that use different percentage of malicious nodes available in the network.

Table 1. Simulation parameters of MANET

Parameters	Description
Routing protocol	AODV
Total nodes	50
Simulation time	100 s
Mac type	802.11
Queue type	Drop tail
Queue length	50
Propagation model	Two way ground
Antenna	Omni
Area	1000 * 1000 m
Radio range	250 m
Malicious rate	10 to 40%
Encryption	AES
Message size	256,512 bytes

Throughput: It is the rate to send packets over the entire transmission range in per unit time, is known as throughput.

$$\textbf{Throughput} = (\textbf{no. of data packets} * \textbf{packet size})/\textbf{total duration in time}.$$

In Fig. 1 comparison has been made between Proposed Throughput & previous Throughput simulation values.

Packet Delivery Ratio: Termed as ratio of delivered data packets at the receiver to packets sent by the sender is known as Packet Delivery Ratio.

$$\mathbf{PDR} = (\mathbf{P_r}/\mathbf{P_s}) * \mathbf{100}$$

Figure 2 shows comparison between proposed PDR & previous PDR.

Delay: Defined as total time taken by the message to travel from sender node to receiver node is termed as delay. It is evaluated as

$$\mathbf{D} = (\mathbf{T_r} - \mathbf{T_s})$$

Figure 3 shows comparison between proposed Delay & previous Delay values.

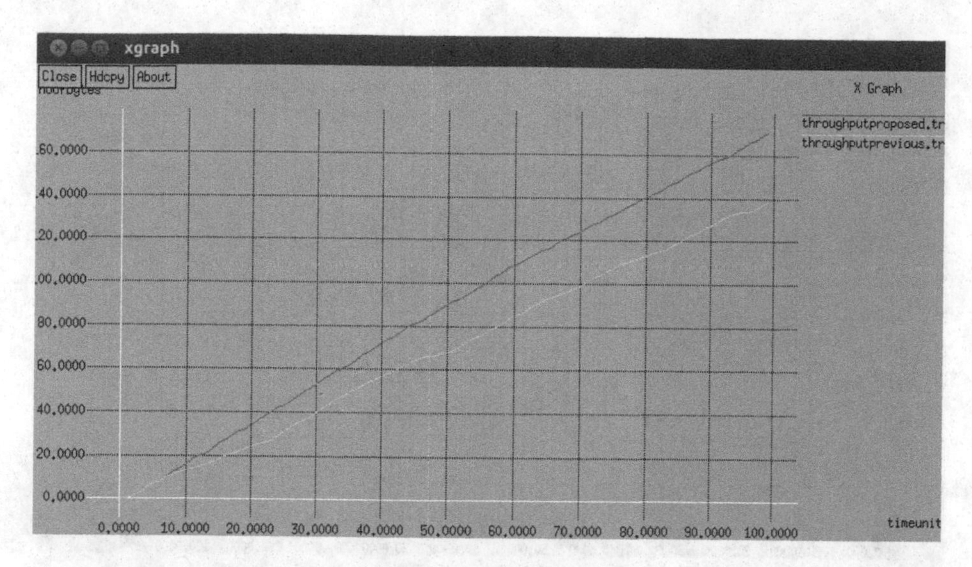

Fig. 1. Throughput

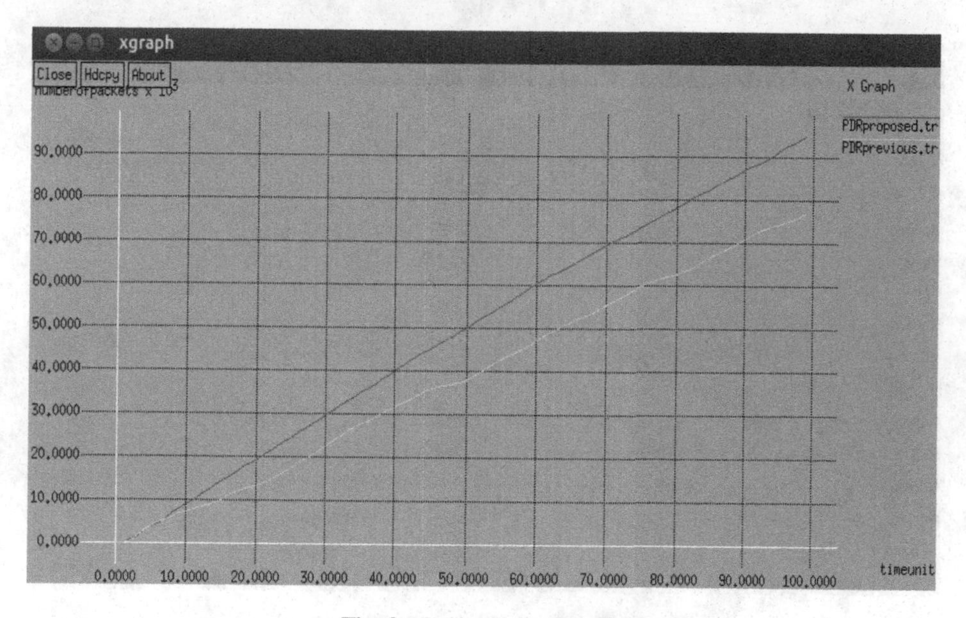

Fig. 2. Packet delivery ratio

Network Overhead: It includes all the routing relating transmissions. For example: RREQ, RREP, REER, ACK etc. Figure 4 represents network overhead occurred over the network due to transmission of various routing packets rather than data packets. Routing packets use high bandwidth of the network that causes degradation in the network. Minimum overhead causes to maximum throughput and packet delivery ratio.

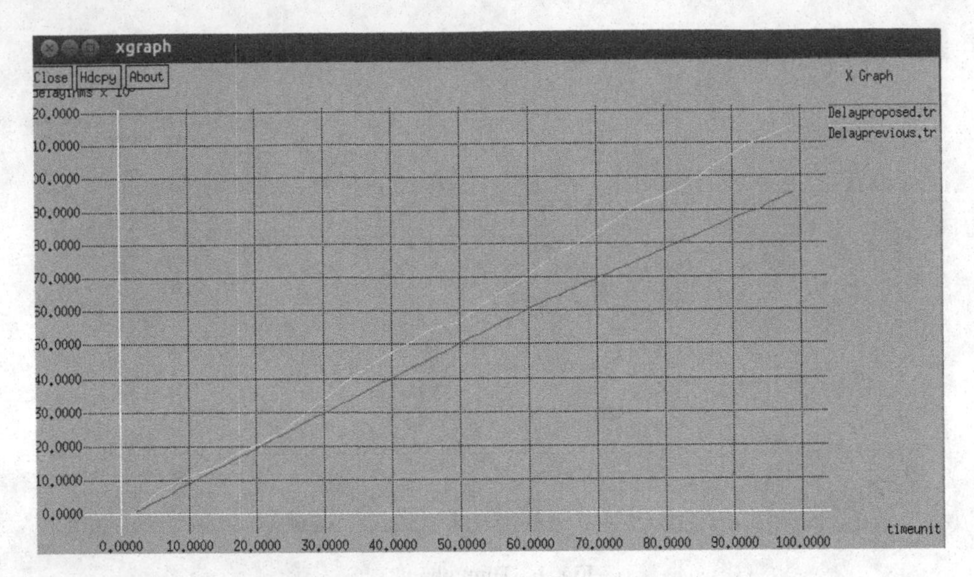

Fig. 3. Delay

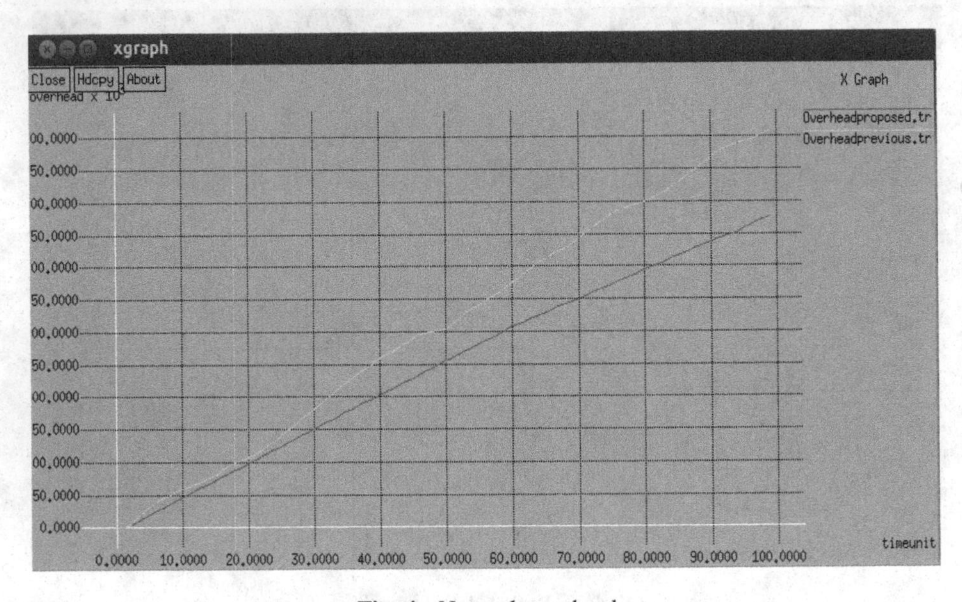

Fig. 4. Network overhead

Figure 5 is used to illustrate the End to End delay over different mobility speed. Data taken at 0, 2, 4, 6… mobility speed. Red line indicates the proposed Delay and Green line indicates the previous Delay.

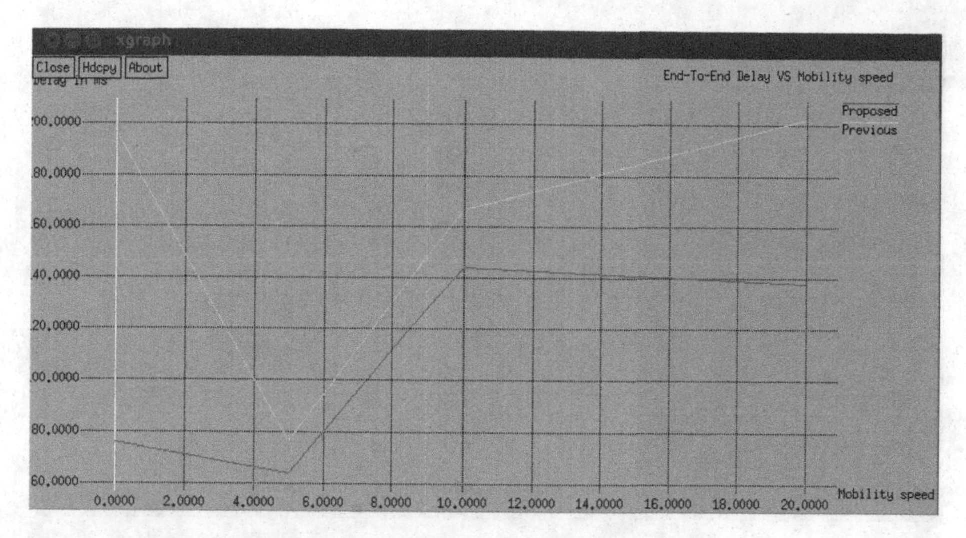

Fig. 5. End to End delay vs mobility speed (Color figure online)

Figure 6 is use to show the Overhead over different modality speed. Data taken at 0, 2, 4, 6… mobility speed. Red line indicates the proposed Overhead & Green line represents the previous Overhead.

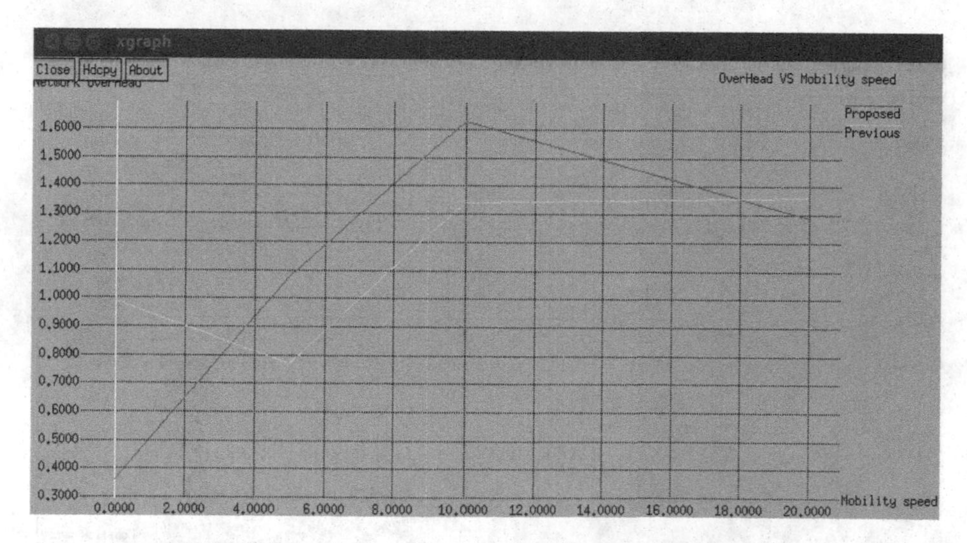

Fig. 6. Overhead vs mobility speed (Color figure online)

Figure 7 is used to represent the PDR over different mobility speed. Data taken at 0, 2, 4, 6… mobility speed. Red line state the proposed PDR & Green line state the previous PDR with speed.

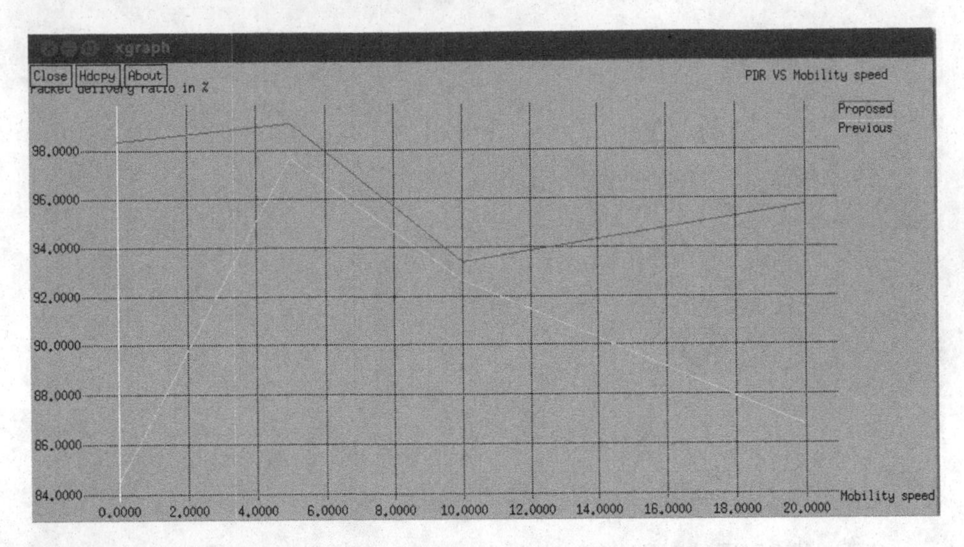

Fig. 7. PDR vs mobility speed (Color figure online)

Figure 8 illustrates the Throughput over different mobility speed. Data taken at 0, 2, 4, 6… mobility speed. Red line states the proposed Throughput & Green line state the previous Throughput values.

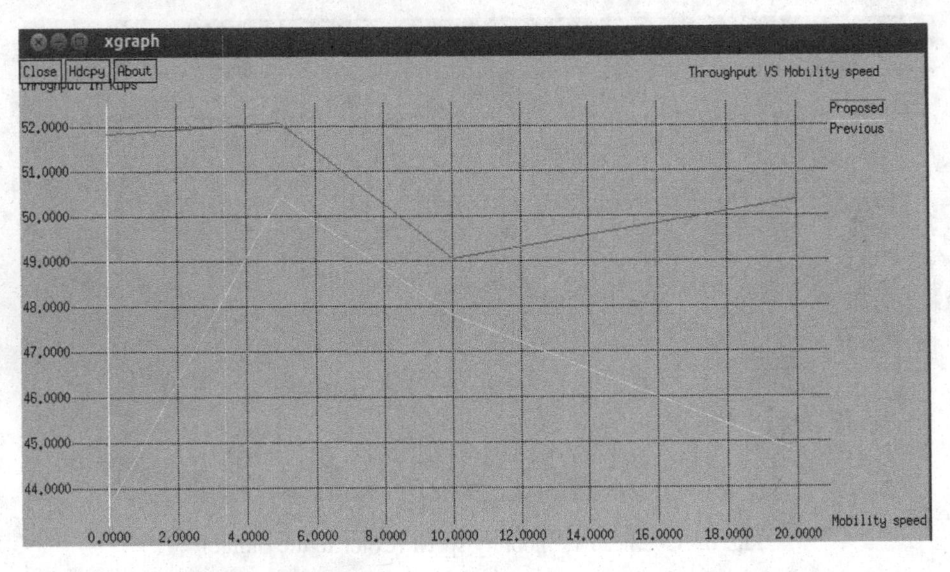

Fig. 8. Throughput vs mobility speed (Color figure online)

In the proposed work different scenarios have been used for simulation of MANET. In this simulation, malicious nodes that have been introduced up to 40% of total nodes over the network.

Table 2 represents values of packet delivery ratio for proposed work under different scenarios of simulation. In the table, comparison has been made between two different approaches of simulation, that are RSA based and AES based encryption. Reverse tracking based mechanism has been used to detect malicious nodes over the network.

Table 2. Packet delivery ratio

Approaches	Malicious: 10%	Malicious: 20%	Malicious: 30%	Malicious: 40%
RSA [19]	84.05	84.11	83.37	80.52
AES (Proposed)	99.02	99.11	98.65	98.80
Overall aggregate % improvement				19.16%

Table 3 represents value of routing overhead for proposed work under different scenarios of simulation. Routing overhead has been caused due to packets that have been transmitted over the network that contain no data but used for route selection and detection of malicious nodes.

Table 3. Routing Over head

Approaches	Malicious: 10%	Malicious: 20%	Malicious: 30%	Malicious: 40%
RSA [19]	3.11	3.27	3.36	3.60
AES (Proposed)	0.96	0.91	1.11	1.19
Overall aggregate % improvement				68.80%

Table 4 represents values of average throughput for proposed work under different scenarios of simulation. Throughput has been measured in the network that has been used to calculate total number of bytes of information has been transmitted over the network with respect to a unit of time.

Table 4. Throughput in kbps

Approaches	Malicious: 10%	Malicious: 20%	Malicious: 30%	Malicious: 40%
RSA [19]	44.67	42.53	41.11	38.62
AES (Proposed)	5.46	49.94	48.47	47.07
Overall aggregate % improvement				14.72%

4 Conclusion

Mobile Ad-hoc Network (MANET) is a set of remote versatile hubs shaping an element self-sufficient system. Hubs speak with one another without the mediation of concentrated access focuses or base stations. Because of its dynamic topology, do not required any central authority. A malicious attacker can rapidly behaves as router and disturbs the system operations intentionally by not following the protocol specifications. Secure communication is an major ingredient of any networking environment, is an especially significant challenge in ad hoc networks. In the previous work a method to find the malicious nodes has been proposed by modifying AODV protocol. In the proposed work AES based encryption approach has been used with AODV routing protocol so that data integrity and confidentiality can be achieved. In the proposed work reverse tracking mechanism has been used for detection of malicious nodes that degrades performance of the network. In the process of reverse tracking mechanism all the nodes that are one hop neighbor from the source node has been identify by transmitting a message and the reply message has been verified if any message has been occurred from other than single hop communication node then the detection mechanism has been enabled for detection of black hole and grey hole attack over the network. In this simulation various performance evaluation parameters have been analyzed that has been used for validation of proposed work. There is overall improvement of 31.17% in throughput, 30.15% in packet delivery ratio, delay has decreased by 12.76% and network overhead has decreased by 17.33%. On the basis of these parameters it can be concluded that proposed approach provide better efficiency and performance as compare to previous one.

References

1. Sanzgiri, K., Dahill, B., Levine, B.N., Shields, C., Royer, E.M.B.: A secure routing protocol for ad hoc networks. In: Proceedings of the 10th IEEE International Conference on Network Protocols (ICNP 2002), pp. 1092–1648 (2002)
2. Papadimitratos, P., Haas, Z.J.: Secure routing for mobile ad hoc networks. In: Proceedings of the SCS Communication Networks and Distributed Systems Modeling and Simulation Conference (CNDS), San Antonio, pp. 27–31 (2002)
3. Hu, Y.C., Perrig, P., Johnson, D.B.: Ariadne: A Secure On-Demand Routing Protocol for Ad Hoc Networks, pp. 21–38. Wireless Networks Business Media, Inc. Manufactured in The Netherlands (2005)
4. Sen, J., Chandra, M.G., Harihara, S.G., Reddy, H., Balamuralidhar, P.: A mechanism for detection of gray hole attack in mobile ad hoc networks. In: International Conference on Information, Communications & Signal Processing, pp. 1–5 (2007)
5. Gao, X.P., Chen, W.: A novel gray hole attack detection scheme for mobile ad-hoc network. In: IFIP International Conference on Network and Parallel Computing, pp. 209–214 (2007)
6. Nadeem, A., Howarth, M.: A generalized intrusion detection & prevention mechanism for securing MANETs. In: IEEE International Conference on Ultra-Modern Telecommunications & Workshops, pp. 1–6 (2009)
7. Cai, J., Yi, P., Chen, J., Wang, Z., Liu, N.: An adaptive approach to detecting black and gray hole attacks in ad hoc network. In: IEEE International Conference on Advanced Information Networking and Applications (2010). ISSN 1550-445X
8. Al-Omari, S.A.K., Sumari, P.: An overview of mobile ad hoc network for the existing protocol and application. J. Appl. Graph Theory Wirel. Ad-Hoc Netw. Sens. Netw. **2**(1), 87–110 (2010)
9. Taggu, A., Taggu, A.: Trace Gray: an application-layer scheme for intrusion detection in MANET using mobile agents. In: IEEE International Conference on Communication Systems and Networks (COMSNETS 2011), Bangalore, India, pp. 1–4 (2011)
10. Patel, M., Sharma, S.: Detection of malicious attack in MANET a behavioral approach. In: IEEE International Advance Computing Conference (IACC), pp. 388–393 (2012)

11. Kanthe, A.M., Simunic, D., Prasad, R.: Effects of malicious attacks in mobile ad–hoc networks. In: IEEE International Conference on Computational Intelligence and Computing Research, pp. 1–5 (2012)
12. Mulert, J.V., Welch, I., Seah, W.K.G.: Security threats and solutions in MANETs- a case study using AODV and SAODV. J. Netw. Comput. Appl. **35**, 1249–1259 (2012)
13. Kalia, N., Munjal, K.: Multiple black hole node attack detection scheme in MANET by modifying AODV protocol. Int. J. Eng. Adv. Technol. **2**(3) (2013). ISSN 2249-8958
14. Sonia, A.A.: A review paper on pooled black hole attack in MANET. Int. J. Adv. Res. Comput. Sci. Softw. Eng. **3**(5) (2013). ISSN 2277-128X
15. Kondaiah, B., Nagendra, D.M.: An efficient protection against collaborative attacks in MANET using cooperative bait detection scheme. Int. J. Innovation Res. Comput. Commun. Eng. (IJIRCCE) **4**(2), 2356–2362 (2016)
16. Bang, A.O., Ramteke, P.L.: MANET: history, challenges and applications. Int. J. Appl. Innov. Eng. Manage. (IJAIEM) **2**(9) (2013). ISSN 2319-4847
17. Vasantha, S.V., Damodaram, D.A.: Bulwark AODV against black hole and gray hole attacks in MANET. In: IEEE International Conference on Computational Intelligence and Computing Research (2015)
18. Goyal, P., Parmar, V., Rishi, R.: MANET: vulnerabilities, challenges, attacks, application. IJCEM Int. J. Comput. Eng. Manage. **11** (2011). ISSN 2230-7893
19. Shakshuki, E.M., Kang, N., Sheltami, T.R.: EAACK- A Secure Intrusion- Detection System for MANETs. IEEE Trans. Industr. Electron. **60**(3), 1089–1098 (2013)

Author Index

Printed in the United States
By Bookmasters

Printed in the United States
By Bookmasters